D0758083

Who Owns Life?

Who Owns Life?

EDITED BY

DAVID MAGNUS
ARTHUR CAPLAN
GLENN McGEE

 Prometheus Books

59 John Glenn Drive
Amherst, New York 14228-2197

Published 2002 by Prometheus Books

Inquiries should be addressed to
Prometheus Books
59 John Glenn Drive
Amherst, New York 14228–2197
VOICE: 716–691–0133, ext. 207
FAX: 716–564–2711
WWW.PROMETHEUSBOOKS.COM

06 05 04 03 02   5 4 3 2 1

Library of Congress Cataloging-in-Publication Data

Who owns life? / edited by David Magnus, Arthur Caplan, Glenn McGee
     p. cm.
  Includes bibliographical references and index.
  ISBN 1–57392–986–7 (alk. paper)
  1. Genes—Patents—Moral and ethical aspects. 2. Genetic engineering—Patents—Moral and ethical aspects. I. Magnus, David, 1961– II. Caplan, Arthur L. III. McGee, Glenn, 1967–

QH445.2 .W46 2002
174'.25—dc21

2002070509

Printed in Canada on acid-free paper

To Julie, Meg, and Monica,
without whom our lives would not be worth owning.

contents

acknowledgments

This book would not have been possible without the help of a lot of people and several institutions. First and foremost, we must thank the Geraldine R. Dodge Foundation for the support they provided. Their generous grant made this book, the conference where a lot of the ideas were first presented, and the Who Owns Life? Web site (www.bioethics.net) possible. We should also thank the many people who attended the meeting and provided input into the papers that were presented there. We also want to thank several other people who worked hard on various stages of this project. First, we want to thank Mark Walters for coming to us to discuss the project and for helping to provide funding. Second, we want to thank Dana Katz and Kay Bradley for helping us to organize the conference and related activities. Third, we want to thank the focus group of high school and college teachers who provided feedback on early stages of the manuscripts, particularly Scott Gilbert. Fourth, we want to thank Stephanie O' Connor and Joe N. Savage Jr. for help with editing, formatting, and generally supporting our efforts. A version of the chapter by Lori Andrews and Dorothy Nelkin is found in *Body Bazaar: The Market for Human Tissue in the Biotechnology Age* by Lori B. Andrews and Dorothy Nelkin (New York: Random House, 2001). The research for the chapter by David Magnus was supported by grants from the Rockefeller Foundation and DuPont, Monsanto, and Dow. Finally, and most important, we thank our spouses, Julie, Meg, and Monica, for letting us play.

who owns life?

INTRODUCTION
DAVID MAGNUS

In some respects, attempts to "improve" on "natural" life-forms have been made for as long as humans have existed. Breeders have utilized the techniques of selection, separation, and crossing to create a bewildering array of forms for a variety of domesticated organisms, from pigeons to dogs to cattle. In the nineteenth and twentieth centuries, breeders came to be a more central part of biological science, producing both important knowledge and new forms of life. Charles Darwin was moved by the results of breeders, and this became the basis for the opening chapter of his seminal *On the Origin of Species*. In the early twentieth century, Luther Burbank became known as "the Edison of the plant world" for the large number of new varieties and species he created and attempted to commercially package. The ownership of each new form came to be an issue during the twentieth century. This came to be especially important as the new science of genetics came to be applied to practical breeding, giving rise to many new forms. Thus, the issue of ownership of life is an enduring one and has been an issue of controversy for over seventy years.

Two forces have recently made this issue of even more central importance to society. First, the pace of technological development has greatly accelerated with the genetic revolution. Celera and the Human Genome Project jointly announced that they have sequenced a draft of the human genome (far more rapidly than previously thought possible) ending a long-standing race between public and private efforts to achieve this milestone. The new sequencing technology will produce vast amounts of sequence data on other organisms as well. This will lead to the discovery of more and more "genes" and the elucidation of some of their functions. The announcement that it was possible to clone an animal (Dolly the sheep) using the nucleus of an adult cell is one of the more vivid examples of the transformations that are taking place. The isolation of human stem cells and a host of patents have followed the development of cloning technology. Genetic engineering has allowed the creation of bacteria that can perform useful functions (such as metabolizing oil from a spill). The patenting of the Chakrabarty bacteria opened the door for the patenting of a host of new engineered organisms—from insect resistant plants, to cows whose milk contain drugs which are useful to humans.

At the same time, a second force accentuated the importance of this issue. Legislative and policy developments in the 1980s have helped to encourage the practice of patenting scientific research (even when done with public funds). These laws and policies are predicated on the idea that the traditional concept of "ownership" has an important role to play in promoting the technological revolution. The Bayh-Dole Act and the Stephenson-Wyler Act (both passed in 1980) helped encourage the patenting of useful inventions or discoveries with the aim of promoting the practical development of products emerging out of scientific research.

This has led to (among other things) an incredible array of patent applications for the discovery of genes—ranging from patents on associations between alleles and diseases, to cDNA and expressed sequence tags (ESTs) for both human and nonhuman genes. Legal challenges of many of the patents are pending, and there is a growing sense of urgency around issues of both gene patenting and engineered forms of life.

These controversies to some extent mirror deeper concerns about the development of these technologies at all. As a result, the "ownership of life" has become one of the most central set of issues facing the new technologies, and may lead us to a host of difficult questions: What is life? What is natural? Do we want to promote the commercial development of these technologies (and when)? Are we somehow "commodifying" life by our practices?

This volume brings together many voices and perspectives on the issue of "owning life." From the patenting of genes and organisms, to the ownership of our bodies and bodily tissues, these are among the most significant moral and social issues facing our society and will form the underpinning of future discussion. When political leaders (including President George W. Bush) discuss their policies about cloning and stem cells, they may have far less power than those who currently hold the patents to nearly all of the underlying technologies.

This volume covers a number of topics and a number of issues. The question of "Who owns life?" is actually a set of interrelated questions. Perhaps the central questions concern the patenting of living organisms. Luther Burbank failed to benefit much financially from the incredible varieties he developed. In the early twentieth century, the desire to encourage agricultural innovation resulted in the 1930 Plant Protection Act. For the first time, it was possible in a significant way to claim the "ownership" over forms of life in the United States, even if only for nonsexually reproducing plants. This gradually came to be expanded to include sexually reproducing plants. The question of whether organisms other than plants could be intellectual property became the subject of a battle as A. M. Chakrabarty pursued a patent on his creation of oil-eating bacteria. The case took many years and finally culminated in the 1980 Supreme Court decision allowing the patenting of an organism that was not a plant, for the first time. Though the Court clearly never envisioned that mammals and other "higher" organisms would be patented, by the end of the decade, a patent had issued on the OncoMouse, a transgenic mouse, opening the door for the patenting of genetically engineered animals, from cows to cats.

From the early plant patents, to the Chakrabarty patent on the oil-eating bacteria, to more recent patents on mammals, the papers by Jack Wilson and A. M. Chakrabarty explore the history of the patenting of living organisms. While Chakrabarty offers his personal experience of pursuing the patent that led to the Supreme Court decision, Wilson provides a greater historical context for that development and begins to raise some of the fundamental issues at stake in allowing these patents to go forward. The article by Rochelle Seide and Carmella Stephens explores the foundations of patent law—what a patent is and how it is applied in the area of biotechnology. Together these three papers provide the background necessary to understand the patenting of life and a limited defense of the current system.

As genetic engineering opened up new avenues for "owning life" so, too, did the technology that made it possible to identify and sequence genes in organisms, including humans. As progress toward the completion of the Human Genome Project took place, patenting of genes became a politically and morally charged issue. Daniel Kevles and Ari Berkowitz argue that the advent of a system that created the race to patent genes and gene fragments is a function of a number of complex factors in the political economy of science. A number of essays explore these developments. Jon Merz argues that disease gene patenting—the patenting of any method of detecting a gene for a disease—is particularly problematic. In at least one case, the families afflicted with a genetic disease helped contribute to the identification of the gene, only to find themselves excluded from decisions about how the test would be developed and utilized. Rebecca Eisenberg explores the question of whether it makes sense to patent genes and in what circumstances. Genes have a strange dual existence as both physical material and informational molecules. Eisenberg argues that it may be appropriate for patents on genes, but only certain classes of patents. This would seem to resonate with Merz's arguments against disease gene patents. The question of whether genes are the sort of things that can be patented is further explored by David Resnik, who goes to the heart of one of the key metaphysical issues that runs through the Wilson, Chakrabarty, Merz, and Eisenberg papers. What counts as an invention as opposed to the discovery of a product of nature? It is well established in U.S. case law that one cannot patent "laws of nature" or "naturally occurring" material. Are genes a part of nature or a creation of humans? Resnik defends the view that genes should be seen as constructs—the product of human invention—and hence patentable material. In his article, Mark Hanson explores a different set of concerns raised by the patenting of both living organisms and genes: Does patenting represent the commodification of life? This is largely the basis of some of the religiously based objections to gene patenting and the patenting of organisms. While the detailed exploration of this topic reveals that most of these sorts of arguments are not terribly persuasive, Hanson argues that they reveal some important and deep-seated concerns that people have with the direction of commercialization of biotechnology.

In addition to genes and organisms, the ownership of life has often been raised over human tissue—the ownership of body parts, bodies, and cell lines. Is there a property right in a person's body? Can a dead person own anything? If not, can someone else own your body? Questions also arise over the ownership of body parts. What happens when an organ is removed? Does the donor retain ownership (at least until it is placed in someone else's body)? By and large, ownership of tissues has been left to the institutions that remove them—hospitals. Are there limits to this? In the Moore case, a man's cells turned out to have valuable properties. For a number of years, researchers stayed in touch with him to procure and patent his valuable cell lines, derived from his blood, bone marrow, skin, and semen. Upon discovery that "a part of him" had been patented, Moore sued the researchers and a litigious battle ensued that led to the California Supreme Court. The 1990s witnessed an increase in the number of "custody" battles over frozen embryos—a combination of the parts of two bodies that arguably leads to the creation of a third. Each of these kinds of cases raises new difficult issues and different ways of framing them—in terms of organ donorship, the disposition of remains, as patentable inventions, as material property, and over custody. They also introduce the specter of the commodification of life through the large sums of money involved in the "body business." The papers by Lori Andrews and Dorothy Nelkin, and Pilar Ossorio explore the entire range of issues introduced by the ownership of the body and its parts and argue that biotechnology represents a sufficient set of challenges, that it is time to recognize that people have a property interest in their bodies.

Biotechnology continues to create new ways to own life. Stem cell research is one of the first major technological developments to take place in the new political economy of biomedical research. Glenn McGee and Elizabeth Banger make the first systematic assessment of the existing stem cell patents and explore the ethical issues they raise. The fact that the intellectual property in this case falls almost entirely in the hands of a small number of relatively small biotech firms could have profound implications for the way research is conducted.

One of the implications of the introduction of patent and ownership rights over organisms, genes, and body parts is that business values have begun to penetrate the laboratory. The Moore cell line was eventually sold for literally millions of dollars. Robert Lee Hotz describes the impact that this has had on the transformation of scientific practice and culture, a change that may not be for the better. The fact that the scientists in the Moore case seem to have lied to him in order to continue their supply of valuable cells is but one indication that scientific values are skewed by financial interests. Lurid tales of theft and competition highlight the underlying reality—commercial interests have transformed the nature of biomedical research. There is increasing pressure on academics to do research that has financial value to their universities, and the race to patent has moved ever further upstream in the development process. At this point, basic science is often patented, and it is not at all clear that this will help with science. Hotz shows how imperative it is that ethical values play a role in the practice of science in the face of huge economic pressures. Meanwhile, in less developed nations, nongovernmental organizations (NGOs) and some governments raise fears that their genetic resources and traditional knowledge of how to use them are being taken away by wealthy industrialized nations and their corporations. David Magnus explores this in his essay. Will the answer to the question "Who owns life?" turn out to be a biotechnology company? And is that a good thing or a problem?

one

PATENTING OF LIFE-FORMS
From a Concept to Reality
A. M. CHAKRABARTY

CONCEPTUAL FRAMEWORK OF AN OIL-EATING PSEUDOMONAD

Back in 1965 after completing my Ph.D. at the University of Calcutta in India, I joined the laboratory of Dr. I. C. Gunsalus at the University of Illinois at Urbana to study the nutritional versatility of a group of microorganisms called *Pseudomonas*. Many members of this genus (collectively called pseudomonads) had the ability to utilize a variety of organic compounds, some as exotic as camphor or naphthalene or various components of crude petroleum. Dr. Gunsalus's interest was to find out how the pseudomonads managed to assimilate so many organic compounds with a genome size only slightly larger than that of *Escherichia coli*, another well-studied microorganism with a limited nutritional ability. My job was to define and map the nature of the bacterial genes that allowed dissimilation of camphor, octane, and similar organic compounds. I was largely able to meet these goals and demonstrated that the genes that allowed the pseudomonads (various strains of *Pseudomonas putida*) to degrade camphor and octane resided not on the chromosome of the bacteria, but on autonomously replicating DNA elements called plasmids. At that time, bacterial plasmids were known to encode resistance to antibiotics or heavy metals and to confer fertility properties to bacteria, but not to enhance nutritional capabilities. Since plasmids are often transmissible from one bacterium to another in absence of the transfer of chromosomal DNA, our studies demonstrated the potential of the pseudomonads to transfer these plasmids (later called degradative or catabolic plasmids) to other bacteria to enhance their range of nutritional versatility.

In 1971 I moved to the Research and Development Center of the General Electric Company (GE) located in Schenectady, New York, as a staff scientist. I was assigned the task of converting cow manure to cattle feed. The manure is basically composed of undigested lignocellulosic materials that are normally resistant to microbial attack. With the help and advice of my colleague, Dr. W. Dexter Bellamy, we developed a strain of thermophilic actinomycetes that could digest the lignin and cellulose components of the manure and grow at their expense, thereby reducing the undigestible lignocellulose content of the manure and enriching it with protein, since the actinomycetes were composed of about 70 percent protein. While this was an interesting and commercially lucrative project, I missed working on the more basic aspects of biology such as an understanding of the nutritional versatility of pseudomonads, the nature and the role of plasmids, and how bacteria acquire the ability to degrade normally recalcitrant compounds.

GE had an intellectually stimulating environment where new ideas were constantly generated and discussed. During a lunch conversation with colleagues, I became aware of the fact that converting crude oil to single-cell protein (a concept very similar to the conversion of lignocellulosic material to bacterial proteins by thermophilic actinomycetes) could be a viable and commercially attractive proposition. In some parts of the world, oil was cheap, but proteins (including fish, animal meat, or plant-derived proteins) were expensive. Conversion of crude oil to bacterial biomass (about 75 to 80 percent protein) could make sense. This discussion immediately kindled my interest since I knew that the pseudomonads were able to degrade various components of crude oil and thus could be an excellent candidate for this bioconversion. However, there was a problem. Crude oil is a mixture of a large number of the hydrocarbons. Individual *Pseudomonas* strains could degrade only a limited number of the hydrocarbon components of crude oil. Growth of the crude oil with a single strain resulted in the loss of a few hydrocarbons and did not allow significant enrichment with bacterial biomass. Different strains, however, could digest different components of crude oil, and thus growth with a mixed culture allowed much more extensive utilization of different hydrocarbons that resulted in significant biomass generation. The problem with mixed-culture growth is that some strains tended to dominate over others, resulting in the survival of the fittest and, of course, overall loss of crude oil conversion. Suddenly, I had an idea. I knew that utilization of some of the crude oil components such as octane, decane, naphthalene, and the like, in *Pseudomonas* was specified by plasmid-borne genes that could be transferred from one bacterium to another. Thus, technically I could construct a multiplasmid single strain that had a number of plasmids, each specifying a different hydrocarbon degradative pathway, so that a single pseudomonad could simultaneously degrade a number of hydrocarbon components of crude oil and grow rapidly, generating significant biomass with high protein content.

This was an attractive idea for me because it allowed me to go back to my postdoctoral days of basic research but at the same time provided me with a rationale for a potential commercial venture. During after-hours and weekends, I started tinkering with *Pseudomonas* genes in an effort to construct a multiplasmid pseudomonad. When some of the degradative plasmids appeared to be incompatible and thus could not coexist, I fused them with UV-irradiation so that they became part of a single, large plasmid. I then checked the ability of the parent strains (from which I transferred various hydrocarbon degradative plasmids) and the newly constructed multiplasmid strain to grow with crude oil as a source of carbon and energy. As expected, the multiplasmid strain grew faster and better with crude oil than the single-plasmid parent strains. This suggested that a genetically engineered pseudomonad with various degradative plasmids could have the potential to generate single-cell protein from crude petroleum in significant amounts compared to natural strains.

THE PATENTING CONTROVERSY: PRODUCT OF NATURE VERSUS COMPOSITION OF MATTER

At about the time the multiplasmid pseudomonad was being constructed, the price of crude oil was rising in the world market. It no longer made much sense to make protein from petroleum because of the high price of petroleum. However, occasional large-scale oil spills from oceangoing tankers in the late sixties and early seventies raised serious concerns regarding the pollution of coastal regions by the spilled oil. Thus the multiplasmid pseudomonad was deemed as a good candidate for oil spill cleanup because of its ability to consume oil relatively quickly. A decision was therefore made at GE to apply for a patent on the process of constructing multiplasmid organisms as well as the organism itself. If this organism was not protected by a patent, anybody could easily isolate it and use it for oil cleanup purposes. I had some discussion with Leo I. MaLossi, the patent attorney working for GE, regarding the latter decision since microorganisms, as living matters, were considered nonpatentable by most patent lawyers working for the pharmaceutical industry. MaLossi could not quite understand why an invention, which may otherwise qualify as a patentable invention, could not be patented simply because it was living. To him, a living microorganism is nothing but a composition of matter and genetic engineering techniques that have imparted a new and useful characteristic to this organism with a consequent change in the composition of its matter. After careful investigation, he decided to include the genetically altered microorganism as a patentable invention in our patent application filed on June 7, 1972.

In late 1973 the U.S. Patent and Trademark Office (PTO) allowed the process claim, but rejected the product claim based on the fact that a microorganism is a product of nature and as such is nonpatentable. In an appeal to the PTO's board of appeals, Leo MaLossi argued that the multiplasmid pseudomonad was not a product of nature since the introduction of the plasmids, including a fusion plasmid generated from incompatible plasmids, changed the characteristics of the strain in a significant way that is very different from other pseudomonads found in nature. The board of appeals accepted this argument but rejected the products claim anyway, arguing that even though the multiplasmid pseudomonad was not a product of nature, it was nevertheless alive and therefore did not merit patent protection.

The rejection of the product claim by the patent office's board of appeals solely because of the living nature of the microbes prompted GE to appeal the decision to the United States Court of Custom and Patent Appeals (CCPA). In early 1978 Judge Giles Rich, speaking on behalf of the three judges in a three-to-two ruling, affirmed the patentability of living microorganisms, thus giving the GE claims a major boost. The CCPA considered microorganisms basically a factory of chemical reactions and various biological transformations and ruled that so long as the microorganism was useful, novel, and a product of human intervention, it merited patent protection irrespective of whether it was living or not.

During this time, another patent application filed by Upjohn Company (on behalf of Malcolm E. Bergy with claims that covered an antibiotic producing *Streptomyces* species) was also considered by the CCPA. In the *Bergy* case, the CCPA also ruled three-to-two, in favor of Bergy. Soon thereafter, the solicitor general of the United States, on behalf of the PTO, appealed to the Supreme Court to rule on the patentability of living microorganisms. The Supreme Court, however, sent the case back to the appeals court for reconsideration in the light of another Supreme Court decision that involved a mathematical algorithm. The CCPA, however, did not see in this decision much relevance to the *Bergy* and *Chakrabarty* cases that involved the question of patentability of life-forms and reaffirmed the patentability of life-forms in 1979; this time ruling by a majority of four-to-one.

THE CASE BEFORE THE SUPREME COURT: PATENTABILITY OF LIFE-FORMS

During the appeal process, the commercial aspects of the recombinant DNA technology were gaining ground, and the proponents of the technology— particularly the pharmaceutical industry, academic institutions, and professional organizations—became deeply interested in the case involving patenting of genetically engineered microorganisms. It was also the time that a great deal of controversy was generated because of the unknown potential hazards of crossing the evolutionary barrier by recombining genes from different kingdoms. Also the PTO was under intense pressure to appeal the CCPA decision to the U.S. Supreme Court. (An excellent account of these activities has been presented by Daniel J. Kevles.)[1] Patent commissioner Sidney Diamond and the solicitor general on behalf of the PTO did just that, and in October of the same year, the Supreme Court agreed to review the case on the patentability of the life-forms. Soon thereafter, however, Upjohn Company withdrew the Bergy product claim, leaving *Diamond* v. *Chakrabarty* as the only case to be decided by the Supreme Court. A number of amicus briefs were submitted to the Court by various organizations, mostly in support of the *Chakrabarty* claim, but a few opposing it. Those who opposed it were contending that life is basically sacred and should not become a subject of the commodity marketplace. Such organizations were also worried that the patenting of a lowly life-form such as a microorganism might set a precedent for the patenting of higher forms of life (perhaps including humans), and that raised the murkiest of issues involving the morality and legality of tinkering with human genes and the ownership of human limbs, organs, or persons.

Once the Supreme Court agreed to take the case, GE hired a renowned attorney, Edward F. McKie Jr., to represent me at the Court. However, by this time, I had accepted a position as a professor of microbiology and immunology at the University of Illinois at Chicago College of Medicine and had moved to Chicago at the end of March 1979. GE offered me a consultantship to confer with McKie, and I talked with him frequently on the scientific definition of life, the biological functions characteristic of living forms, and how genetic techniques, particularly viral genome sequences and characterization of gene products, were enabling scientists to define precisely how various living functions involved a series of chemical and biological reactions. Such understandings were at the core of Judge Giles Rich's argument that there is nothing unique to the living form that can make a microorganism nonpatentable simply because it is living. In addition to his law degree, Ed McKie had a degree in engineering and used to pose sharp questions for me involving the definition (or as near a definition as one could come up with) of life or what may constitute life. In March 1980 the Supreme Court justices heard oral arguments of this case, which I was privileged to attend. The arguments centered on whether the techniques used in the construction of the oil-eating pseudomonad were or were not completely new technologies, whether they may pose unprecedented or unforeseen problems, and why patents could be issued to new chemicals but not to new bacteria even though such bacteria represented nothing more than an altered composition of matter. The broad questions of patentability of higher forms of life were set aside, and the arguments were focused on the narrow definition of the patentability of an altered form of a soulless, mindless, and lowly form of life.

On June 16, 1980, the Supreme Court delivered its verdict. In a close five-to-four decision, the Court held that the multiplasmid, oil-eating pseudomonad was not a product of nature but a man-made invention that merited patent protection. The Court brushed aside many of the scenarios that predicted calamitous consequences for allowing the patenting of life-forms. The justices affirmed that their decision was a narrow one dealing with the subject in question and not addressing bigger questions of patenting higher forms of life, which was the subject of a congressional debate and action.

BEYOND THE SUPREME COURT DECISION: PATENTING OF HIGHER LIFE-FORMS

Even though the Supreme Court based its decision in a focused manner centered on a genetically engineered bacterium, the PTO interpreted the decision in a much broader manner, granting patents on genetically altered plants, animals, human cells and tissues, disease genes, and the like. I have reviewed this area previously and will not dwell on it in detail, except to point out that granting patents on animals, birds, plants, fish, and human cells and genes (including expressed sequence tags [ESTs] and single nucleotide polymorphisms [SNPs]) represents an outcome wholly unforeseen, but to some extent anticipated or feared during the controversy surrounding the patenting of the oil-eating pseudomonad.[2] Publication of the complete sequence of the human genome will provide important insights into the role of genes not only in the causation of disease, but also in various other facets of human life, including human behavior, aging, emotional status, or susceptibility to infectious diseases or environmental insults. With progress in such areas will come the attendant ethical, moral, and legal issues, including the issues on intellectual property rights to human organs made in vitro by triggering differentiation of cultured embryonic stem cells or to hypothetical human-animal hybrids.[3] Since such issues invariably end up in a court of law, there are efforts underway to initiate a dialogue between the members of the scientific community and the judiciary so that all aspects of the science and law, particularly those involving human genetic intervention, are understood and debated in the courts. Thus, the Supreme Court decision on *Diamond* v. *Chakrabarty* appears to have gone beyond what the Supreme Court justices perhaps intended to grant. The subject of "who owns life?" has therefore become a significant, timely, and dominant issue of our times.

NOTES

1. Daniel J. Kevles, "Ananda Chakrabarty Wins a Patent: Biotechnology, Law, and Society, 1972–1980," *Historical Studies in the Physical and Biological Sciences* 25 (1994): 111–35.

2. Ananda M. Chakrabarty, "*Diamond* v. *Chakrabarty*: A Historical Perspec-

tive," in *Principles of Patent Law*, Donald S. Chisum, Craig A. Nard, Herbert F. Schwartz, Pauline Newman, and F. Scott Kieff (New York: Foundation Press, 1998) 783–88.

3. Rick Weiss, "Patent Sought on Making of Part-Human Creatures: Scientist Seeks to Touch Off Ethics Debate," *Washington Post*, 2 April 1998, A12; *Judges Journal* 36 (1997): 1–96.

two

PATENTING ORGANISMS
Intellectual Property Law Meets Biology
JACK WILSON

W hile Ananda Chakrabarty worked for General Electric, he bred a strain of the bacterium *Pseudomonas* that could degrade crude oil more efficiently than the multiple bacterial strains used for the same purpose. His bacterium had several advantages over the mixed cultures—it contained extra plasmids, each of which could degrade a different component of crude oil, and it was not subject to the mutual inhibition or differential reproductive rates that made bacterial mixtures less effective.

Chakrabarty applied for a patent on his *Psuedomonas* strain in 1972. Despite some popular confusion on this point, he did not use recombinant DNA techniques to alter the bacteria with which he started—those methods hadn't been invented yet. By the time Chakrabarty's case had been decided in the Supreme Court, nothing short of a revolution had occurred in biologists' abilities to splice together genetic material from different organisms. From 1972 to 1980 molecular biologists developed a series of techniques that made it possible to reliably splice DNA from one organism to another, creating hybrids that were previously impossible to make. The scientists who developed these methods had at least applied for patents on them before the *Chakrabarty* case was decided.

Chakrabarty's patent application, denied first by the patent examiner and then by the U.S. Patent Office Board of Appeals (POBA) in 1976, traveled through several legislative offices and judicial channels before being decided in the Supreme Court in June 1980. The application left in its wake a series of convoluted and contradictory interpretations of patent law. The *Chakrabarty* case not only raised questions about the patentability of a biological organism, but it exemplified a transformation in the purpose and interpretation of patent law that took place throughout the twentieth century, a shift from primarily granting patents on conventional mechanical inventions to patenting chemicals, plants and animals, and even genes.

A patent is an intellectual property right granted by state authority that excludes others from the use or benefit of a patented invention without the consent of the patentee. In the United States, a patent applicant has to meet the following four requirements: novelty, nonobviousness, usefulness, and adequate specification/disclosure. An applicant may try to patent either a *process* for making a product or the *product* itself (regardless of how it is produced). According to the Constitution, "Whoever invents or discovers any new and useful process, machine, manufacture, or composition of matter or any new and useful improvements thereof, may obtain a patent therefore subject to the conditions and requirements of this title" (U.S. Code Title 35 Section 101). The Constitution clearly states that the patent system was established in order to "promote the Progress of Science and the useful Arts by securing for limited Times to Authors and Inventors the exclusive Right to their respective Writings and Discoveries" (art. 1 sec. 8). Unlike some European patent systems, the U.S. law finds its justification in its instrumental power to promote scientific progress, not in a basic right to one's own intellectual property.

A U.S. patent grants the patent holder the right to exclude others from producing the patented invention for twenty years from the time of application. The law offers two incentives to would-be inventors. It encourages them to invest time, money, and effort in projects that would be unprofitable to pursue without patent protection, and it encourages them to disclose inventions rather than keep them secret. A scientist can invest resources into the development of a new drug, for example, that is expensive to discover or test but inexpensive to manufacture once developed. In return for a twenty-year monopoly on that invention, the inventor discloses how the invention is made to the extent that it can be replicated by one "skilled in the art."[1] The patent encourages disclosure of information that otherwise might not be open to scrutiny by other scientists or business competitors— in biotechnology these roles may be played by the same people.[2] Trade secrecy does not protect the inventor against reverse engineering—it lacks the blocking power a patent has. Legally maintaining trade secrecy requires abstaining from publishing your findings. Secrecy also violates the Mertonian norms of science—it short-circuits the normal rewards of peer recognition and open communication. Also, if the invention is likely to be independently discovered by someone else, it pays to claim it as soon as possible—biotechnology is just such a field. "The field of genetic research was so competitive that if one didn't publish findings quickly (often within a matter of weeks or months), there was a constant prospect that some other group would."[3] Though the businessperson's interest can often be in conflict with the interest of the academic scientist, in biotechnology those interests can be complementary. The "patent and publish" strategy ensures protection for the invention and, by changing the state of the "prior art" (existing body of knowledge), blocks other patents. If you *first* publish your findings and *then* seek patent protection, you add your paper to the prior art and may negatively affect your own claim on a patent.

When Chakrabarty applied for a patent, he claimed "a bacterium from the genus *Pseudomonas* containing therein at least two stable energy-generating plasmids, each of said plasmids providing a separate hydrocarbon degradative pathway," as well as a process using the bacteria. The patent examiner accepted his claim for a *process* that included the use of straw inoculated with the new bacteria to contain oil spills. Patents for processes involving microorganisms were not unprecedented—several septic system patents had claimed bacterial processes as subject matter. However, the examiner rejected on two grounds Chakrabarty's claim for a patent on the organism itself—maintaining that the microorganisms were "products of nature" and they were drawn to "live organisms." The examiner referred to the Plant Protection Act of 1930 to argue that patent law was not intended by Congress to include living organisms.[4]

Before the Plant Protection Act (PPA) was passed in 1930, plants were not considered to be patentable subject matter. The PPA extended patent protection for unique plants discovered or invented in a cultivated state, provided that they could be reproduced asexually. A plant patent excludes anyone but the patent holder from asexually reproducing the patented plant. The examiner in Chakrabarty's case invoked the PPA to support his conclusion that because they were alive and they were not plants, the bacteria were not patentable. His interpretation precipitated a series of conflicted exegeses of congressional intent in passing the PPA.

Chakrabarty appealed the examiner's rejection to the patent office board of appeals (POBA). The POBA affirmed the examiner's initial rejection of the claim, but did so exclusively on the grounds that living organisms, even if they were sufficiently modified so as not to be products of nature, were not patentable subject matter under U.S. Code Title 35 sec. 101. In doing so, the POBA dropped the initial examiner's assertion that the claimed bacteria were "products of nature." The board reversed the examiner on this point, agreeing with Chakrabarty that the claimed bacteria were not naturally occurring.

Chakrabarty appealed the patent office board's rejection to the U.S. Court of Customs and Patent Appeals (CCPA) where it was heard by a three-judge panel that reversed the examiner and the POBA's decision on the grounds that it was not relevant fact that the subject matter in Chakrabarty's claim was alive, so long as it met the general conditions for patentability.

> In the instant appeal, appellants are seeking protection for a new bacterium, admittedly alive, in which such changes have been effected as to produce in this bacterium new capabilities. The Board of Appeals has agreed that this organism is not a "product of nature." If it be accepted that all things in our world are either products of nature or things produced by man, then by the process of elimination the Board of Appeals has agreed with the appellant's contention that his new bacterium is a thing produced by man, i.e., a manufacture. It should follow, therefore that appellant has already met the requirements of Section 101.[5]

Chakrabarty's case, as it traveled from the POBA to the CCPA, focused not on whether Chakrabarty's bacteria were useful, but almost exclusively on whether or not living organisms other than some asexually reproducing plants (already excepted by the 1930 Plant Protection Act) could be subject to patent protection. The issue as to whether the bacteria were products of nature or man-made products had been dropped by the POBA, so whether or not living organisms other than asexually produced varieties of some plants were subject to patent protection seemed, to the CCPA panel, to be the only issue on which to rule. Judge Howard Markey, for example, argued that

> The sole issue before us is whether a manmade invention, admittedly novel, useful, and un-obvious, is unpatentable because, and only because, it is "alive" (in the sense that microorganisms are "alive").

The CCPA overturned the POBA's decision, ruling that the fact that the new bacterium was alive did not disqualify it from being subject to patent law. They upheld Chakrabarty's claim and their decision, in effect, implied that a living organism should now be categorized and protected by the Constitution, as "a composition of matter or manufacture."

The CCPA's decision was not unanimous. Judge Phillip Baldwin dissented, "I find no admission by anyone that the present invention is a statutory 'manufacture.' 'Manufacture' and 'manmade' are not synonymous for patent purposes." Baldwin cited *American Fruit Growers* v. *Brogdex Co.*, heard before the U.S. Supreme Court in 1931. The Supreme Court ruled in this case that although boraxed oranges do not occur in nature, they are not patentable because oranges are not statutory manufactures. This category of natural objects modified by human effort fell ambiguously between manufacture and product of nature. Judge Baldwin asserted that

> The law, as propounded by the Supreme Court, defines three alternatives. Between true "products of nature" and statutory subject matter or "manufactures" lies an intermediate category of things sufficiently modified so as not to be products of nature, but not sufficiently modified so as to be statutory "manufactures." Therein are found the borax-impregnated oranges of American Fruit . . . and in my view, the organisms now before us.

American Fruit Growers v. *Brodgex Co.* provides a commonsense precedent that not all modifications of natural objects are sufficient to transmute a product of nature into a statutory manufacture. But it also leads to the questions about what degree or nature of modification is necessary to transform an admittedly unpatentable living thing into a statutory subject. Baldwin's criterion for this distinction is an odd one:

> I read *American Fruit* as saying that a modified natural product does not become statutory subject matter until its essential nature has been substantially altered. . . . Applying the *American Fruit* rule to the modification of living organisms and to the case before us, I believe that the essential nature of the unpatentable organism with which the applicant started was its *animateness or life* [emphasis added].

Perhaps, by his standard, killing an organism and thus eliminating its animateness would be enough to make it statutory subject matter. Judge Baldwin goes on to argue that Chakrabarty has not changed the essential nature of the bacteria; he only made it better at digesting oil, putting the bacteria in the same category as the boraxed oranges from American Fruit—too modified to be a product of nature, but insufficiently modified to have been changed in its essential properties or rendered statutory manufacture. As more than two thousand years of philosophical disagreement have made clear, referring to essential natures is unlikely to provide a clear standard for distinguishing between these three categories.

Judge Jack Miller, who also dissented with the CCPA's decision but on more concrete grounds, cited the Plant Protection Act of 1930 and argued that Congress thought it was extending a new privilege to plant breeders. The congressional reports leading up to the act make clear the fact that it was being treated as a new extension of patent law rather than as a reapplication of an existing statute. Miller argues that we must presume new legislation is not superfluous:

> Thus, the legislative history clearly shows Congressional understanding that, under the patent law in effect prior to the Plant Act of 1930, reward for service to the public in developing new varieties of plants had not been extended to inventors.

Judge Miller reasons that if we presume, as we must, that the Plant Protection Act was not superfluous, then we must conclude that it was the intent of Congress to not include living things other than new varieties of plants as patentable subject matter. The only other interpretation of congressional intention is to claim that the PPA was enacted primarily to relax the disclosure rules, which at that time required a full written disclosure of the invention. The PPA relaxed the disclosure requirement allowing asexually reproduced plants be described "as far as possible."

The Supreme Court remanded Chakrabarty's case back to the CCPA for "further consideration in light of *Parker* v. *Flook*," in which a mathematical formula was held to be unpatentable. The CCPA maintained that *Parker* v. *Flook* was irrelevant and affirmed its original opinion. Judge Baldwin reversed his initial opinion and concurred with the panel while Judge Miller wrote a dissenting opinion that resembled his original one.[6]

The commissioner of patents appealed the CCPA's second decision, which reaffirmed their first, to the Supreme Court (*Diamond, Commissioner of Patents and Trademarks* v. *Chakrabarty*), where approximately the same argument took place resulting in a five-to-four decision that living things are patentable. In short, the majority argued that the patent statute should be interpreted broadly so that manufacture and composition of matter include "anything under the sun that is made by man," including living organisms.

The limitations of "anything made by man," the criteria for invention, had been established in part by *Funk Brothers Seed Co.* v. *Kalo Inoculant* (1948), where it was decided that combining bacteria in a single package does not sufficiently constitute a manufacture or composition:

> Each of the species of root-nodule bacteria contained in the package infects the same group of leguminous plants which it has always infected. No species acquires a different use. The combination of species produces no new bacteria, no changes in the six species of bacteria, and no enlargement of the range of their utility. Each species has the same effect it always had. The bacteria perform in their natural way. Their use in combination does not improve in any way their natural functioning. They serve the ends nature originally provided and act quite independently of any effort of the patentee.[7]

The opinion in *Funk Brothers Seed* was not unanimous, but the patent was overturned by a seven-to-two margin. Justice William O. Douglas wrote the majority opinion and held that the patent was invalid because the mixture of several species of bacteria was not an invention, merely the packaging of an innovative discovery—the discovery that several species of *Rhizobium* were mutually compatible. Though the *Funk Bros.* decision has been used to help clarify the boundary between products of nature and patentable inventions, it is worth noting that Justice Felix Frankfurter's concurring opinion was based not on the absence or lack of invention but on the absence or lack of clarity in the scope of the patentee's claims. Indeed, Frankfurter did not find Douglas's position compelling:

> It only confuses the issue, however, to introduce such terms as "the works of nature" and "the laws of nature." For these are vague and malleable terms infected with too much ambiguity and equivocation. Everything that happens may be deemed "the work of nature," and any patentable composite exemplifies in its properties "the laws of nature." Arguments drawn from such terms for ascertaining patentability could fairly be employed to challenge almost every patent. On the other hand, the suggestion that "if there is to be invention from such a discovery, it must

come from the application of the law of nature to a new and useful end" may readily validate [the patent holder's] claim.

The *Funk Brothers Seed* decision is frequently credited with and cited for having marked the boundary between products of nature and patentable inventions, but it does not seem to offer a clear or reliable standard.

When the Supreme Court heard Chakrabarty's case, it revisited *Funk Brothers Seed* to conclude, "It would seem that the claim in *Funk Brothers Seed* was not in fact for a true product of nature. The claimed mixed culture did not exist in natural form. *The Funk Brothers Seed* decision is perhaps best viewed as an interpretation of the 'nonobviousness' or 'invention' requirement, and not of the statutory classes of subject matter." The Court found that the Chakrabarty bacterium was a new one with new features neatly assembled within a single cell membrane—they did not apply the product of nature disqualification, a disqualification that seems to be invoked or applied arbitrarily or as a cover for some other problem, that is, novelty or obviousness.

The Supreme Court rejected dissenting arguments from members of the CCPA panel who argued that Congress did not intend to extend patent protection to organisms *except* for asexual plants. "We reject this argument. Prior to 1930, two factors were thought to remove plants from patent protection. The first was the belief that plants, even those artificially bred, were products of nature for the purposes of patent law." This first belief was documented in *Ex parte Latimer*, 1889 December Com. Pat. 123, when the Court denied a patent on a natural fiber found in a pine needle. The Plant Protection Act (PPA) of 1930 made, for the first time, a "clear and logical distinction between the discovery of a new variety of plant and of certain inanimate things, such, for example, as a new and useful natural mineral." Congressional discussants noted that "the mineral is created wholly by nature unassisted by man [while] a plant discovery resulting from cultivation is unique, isolated, and not repeated by nature, nor can it be reproduced by nature unaided by man." When Congress passed the Plant Patent Act in 1930, it—not the courts—set the limits of patentability. And it employs a wide range of latitude when setting these limits. The Supreme Court has made a good case for steering clear of the moral issues involved however real they may be.

It has been argued that "Whatever their validity, the contentions now pressed on us should be addressed to the political branches of the government, the Congress, and the executive, not the courts." Many bills have been sponsored but none have passed. The Court maintained that its task, in deciding *Chakrabarty*, was "the rather narrow one of determining what Congress meant by the words it used in the [Plant Protection] statute."

Justice William Brennan dissented, arguing that it was not the role of the courts to extend the law if the Court admits that the popular conception in 1930 was that plants were not patentable.

> Thus, we are not dealing—as the court would have it—with the routine problem of "unanticipated inventions." In these two Acts [Plant Protection Act of 1930 and Plant Variety Protection Act of 1970], Congress had addressed the general problem of patenting animate inventions and has chosen carefully limited language granting protection to some kinds of discoveries, but specifically excluding others. These Acts strongly evidence a congressional limitation that excludes bacteria from patentability.

According to Brennan, the Court must explain why, if not to specifically exclude certain discoveries, the acts were enacted in the first place. If we presume the acts were not idle exercises or mere corrections of the public record, it seems clear to him that Congress intended, through these acts, to create a subset of animate objects, plants, as patentable while excluding other innumerable animate patents.[8]

THE 1930 PLANT PROTECTION ACT

The 1930 Plant Protection Act anticipated many of the issues that later arose in the *Chakrabarty* case; the legitimacy of the *Chakrabarty* decision depended in part on the interpretation of congressional intent in 1930.[9] Prominent congressmen and politicians of the day had long since championed legislation to protect plant breeders' intellectual property. Paul Stark, chairman of the National Committee for Plant Patents, for example, promoted legislation in Congress, and Thomas Edison supported this legislation with a telegram. Congressman Fiorello La Guardia (later mayor of New York), initially tried to block the new patent legislation but later changed his mind. Among the evidence submitted in favor of protection for plant inventions was a posthumous letter written by Luther Burbank, the famous plant breeder.

A man can patent a mousetrap or copyright a nasty song, but if he gives
the world a new fruit that will add millions to the value of earth's annual
harvests he will be fortunate if he is rewarded so much as having his
name connected with the result. Though the surface of plant experimentation has thus far been only scratched and there is so much immea-
surably important work waiting to be done in this line I would hesitate
to advise a young man, no matter how gifted or devoted, to adopt plant
breeding as a life work until America takes some action to protect his
unquestioned right to some benefit from his achievements.[10]

The Plant Protection Act (PPA) seems to have been a muddled piece
of legislation from the beginning and was clearly perceived at the time as
an extension of contemporary patent law.

Plant patents had not been granted prior to the PPA for two reasons.
First, it was generally believed that even after selective breeding, plants were
products of nature (*ex parte Lattimer*) and second, the Court maintained that
plant inventions could not be adequately disclosed in a written description.
Both of these concerns were addressed in Congress—the law was rewritten
so that the description of the plant only had to be "as complete as reasonably possible." In a letter to the senate committee, Arthur Hyde, secretary of
agriculture, explained that plants are to be brought under patent protection
that, at the time, was only extended to nonliving things:

> This purpose is sought to be accomplished by bringing the reproduction
> of such newly bred or found plants under the patent laws which at the
> present time are understood to cover only inventions or discoveries in
> the field of inanimate nature.

Newly discovered or bred plants were to be protected as having been
developed "in aid of nature," and they were to be described in as much
detail as was reasonably possible. Patent legislation was changed to include
"Any person who has invented or discovered any new and useful
machine . . . or who invented, discovered, and asexually reproduced any
distinct and new variety of plant other than a tuber-propagated plant"[11]
The plant, according to section 4886 of the revised statute, had to be found
in a "cultivated state" and had to be asexually reproduced.

The Supreme Court in *Chakrabarty* interpreted the PPA as if the legislation draws a distinction between *products of nature* and *manufactures* rather

than between *living things* and *nonliving things*. But in passing the PPA, Congress does not seem to have had the former distinction in mind. Not only did the PPA not cover all plant inventions (e.g., tuberous plants), but it did cover things that were clearly not manufactures—found mutations, for example. Indeed few of the first plant patents involved manufacture or the active intervention of the patentee. The asexual reproduction requirement was an easy one to meet, and not sufficient to distinguish the patentee as having done something noteworthy. Moreover, the PPA was enacted to "afford agriculture, so far as practicable, the same opportunity to participate in the benefits of the patent system as [had] been given industry, and thus assist in placing agriculture on a basis of economic equality with industry."[12] Congress sought to "remove existing discrimination between plant developers and industrial inventors" and does not seem to have differentiated products of nature from statutory manufactures. [13]

In eighteen of the first sixty plant patent applications, for example, the applicants stated that their new plants were sports (abnormal growths on a parent plant), but the patents contained no information regarding how the sports were produced. In fact, in most cases, it does not appear that the applicant did anything but find the sport and then reproduce it in the usual way.[14] The law, as written, prevents patenting plants discovered in nature but allows patenting of plants found in a cultivated state, even if *nothing was done to produce them*. Plant patent 25, for example, was for a hybrid tea rose, "a sport on a Talisman rose found at Cromwell, Connecticut." Breeders may have acted actively, if ignorantly, in the process through irradiation, hand pollination, or the use of chemical mutagens, but then again, they may not have.

After the PPA was passed, *Heredity*, a journal otherwise embroiled in the eugenics debate, published a series of derisive articles. When the first plant patent was granted to Henry Bosenberg, a landscape gardener, an article in the journal reported that he "bought a number of Van Fleet roses for use in his work. One of these proved to be a bud sport, apparently, and a new variety had been—'invented'! The new rose is exactly the same as the well-known climbing rose originated by Dr. Walter Van Fleet, and bearing his name, except that it is claimed to be everblooming. . . . The owner of a dozen or so vaguely defined patent claims can, if he has the right psychology and money enough to hire a lawyer, cause almost complete cessation of improvement of a given plant."[15]

The right accorded by the PPA seems to have been the right to exclude others from asexually reproducing the patentee's plant.[16] The patent applicant had to be able to asexually reproduce the plant, but it was difficult to determine if the patent had been infringed upon because plant production could not be wholly described or disclosed. In 1952 the Plant Protection Act was revised to make it clear that the patent holder had the right to prevent anyone else from asexually reproducing or selling unauthorized reproductions of the plant. But the act did not address the problem of patenting newly discovered products of nature—the fact that finding a plant, recognizing that it is novel, and asexually reproducing it are sufficient grounds for patent protection, that nothing need be altered, and that no work has to be done. The plant must be found in a cultivated state, but the discovery does not necessarily entail any creative work.

Congress passed the Plant Variety Protection Act (PVPA) in 1970 to provide patent-like protection to new, distinct, uniform, and stable varieties of plants that reproduce sexually.[17] The PVPA did not technically allow for patent protection, but offered patent-like protection for sexually reproducing plants. The current law, revised in 1994, extends protection for twenty years (twenty-five for trees or vines) just like a patent, but is administered by a different government agency. The PVPA included research and farm exemptions absent from the PPA. The scope of protection a plant patent provided was not clearly defined until 1995. The term "variety" as in "the discovery of a distinct and new variety" was applied and interpreted differently in the PVPA than in the PPA.

IMMEDIATE EFFECTS OF *CHAKRABARTY* PRECEDENT

The *Chakrabarty* case set a precedent that soon changed how patent law was applied to biotechnology, but curiously did not effect a literal change in the law. Important decisions have all been patent office policy; though several attempts have been made, no relevant legislation has made it through Congress. Although Chakrabarty's bacterium was not created using recombinant DNA techniques, by the time his case was decided in 1980, nearly the complete set of recombinant techniques had been invented/discovered, and some popular reports actually clouded the details of Chakrabarty's case. The first biotechnology companies had been founded, and there were a number of patent applications waiting to be processed.

In *ex parte Hibberd*, 1985, the patent office board of appeals (POBA) ruled that plants can be patented without following the special provisions of the PPA or the PVPA. This was an administrative decision within the U.S. Patent and Trademark Office and is not binding on the courts. Their decision was based on *Chakrabarty*.

In *ex parte Allen*, 1987, the POBA ruled that a multicellular animal (a polyploid oyster) was patentable subject matter, again citing *Chakrabarty* and also *Hibberd*.

> The issue . . . in determining whether the claimed subject matter is patentable under Section 101 is simply whether that subject matter is made by man. If the claimed subject matter occurs naturally, it is not patentable subject matter under Section 101. The fact, as urged by the examiner, that the oysters produced by the claimed method are "controlled by the laws of nature" does not address the issue of whether the subject matter is a non-naturally occurring manufacture or composition of matter.[18]

The oyster was held to be a nonnaturally occurring manufacture of matter within the meaning of U.S. Code Title 35 Section 101, but the patent was rejected because it was ruled to be obvious and therefore not patentable.[19]

After *ex parte Allen*, Donald J. Quigg, assistant secretary and commissioner of patents and trademarks, made a statement in a news release on April 4, 1987, that "the Patent and Trademark Office now considers non-naturally occurring non-human multicellular living organisms, including animals, to be patentable subject matter within the scope of 35 USC 101." Quigg referred directly to *Diamond* v. *Chakrabarty* in his explanation. He explained that in order for living organisms, including animals, to be patentable, "they must be given a new form, quality, properties, or combinations not present in the original article existing in nature." No property rights could be held in human beings because property rights in humans had been prohibited in light of antislavery laws.[20] None of the post-*Chakrabarty* decisions mentioned above went to court; all of them were settled by the patent and trademark office in light of the *Chakrabarty* decision.

The first patent on a multicellular animal was granted to Harvard University scientists in 1988. The patent on the Harvard "OncoMouse" was

licensed to DuPont, and these mice were sold to cancer research laboratories. The mice contain several genetic sequences containing oncogenes and
promoter regions that make them susceptible to developing tumors. The
actual patent (No. 04,736,866) is for "a transgenic non-human mammal,
all of whose germ cells and somatic cells contain a recombinant activated
oncogene sequence introduced into said mammal, or an ancestor of said
mammal, at an embryonic stage." The mouse can be purchased from
DuPont for approximately $50.

As of 1995 very few multicellular animal patents—most of them on
mice—had been issued and, at first, the entire Chakrabarty affair seemed
to have been a tempest in a teapot. In the years following *Chakrabarty*,
there seemed to be "little evidence of invention relating to higher animals
that can be eaten, milked, plucked, or shorn for human advantage" and
"technological interest in higher animals [seemed to] end at the level of
the rodent."[21] But as of last year, approximately 1,500 patent applications
on a wide variety of multicellular animals had been filed covering everything from seafood to cattle that express human hormones in their milk.

ANALYSIS OF THE STATUS QUO

In this section I will sketch how the intellectual property rights regime
developed by the U.S. patent office in light of the *Chakrabarty* decision has
been applied to biological innovations once it was established that organisms are patentable subject matter. I will describe some of the persistent
problems that have arisen from these developments and suggest improvements in the patent office.

Some of the controversy surrounding organism patents has been less
about objections to the *patenting* of organisms and more the result of general antipathy toward biotechnology in general, or the processes by which
organisms are modified and used for human ends. Concern about whether
it is morally acceptable to alter an animal or plant's genetic material are not
going to be my main focus because they are not relevant to the patent law
as it is written or as it is applied. A patent seeker need not establish that his
invention promotes the good of humanity, or that its use would be morally
justified. There are two reasons why this is so. First, no morality requirement is written into the general patent law, and second, following the New

Patent Act of 1952, the law is clear that a patent brings with it no explicit or implicit right to make or use the patented invention, only the right to prevent others from doing so without the patent holder's consent.[22] In fact, an effective way to block the use of a technology would be to patent it and refuse to license it. By taking this position, I am not arguing that biotechnology does not raise important moral issues, but rather that most of these are distinct from questions about patenting. There may be good reasons to be morally uneasy about stem cell research, but that does not automatically make it a patent issue. Because this is so, moral concerns about the modified organism itself, rather than about the patenting of such organisms, is beyond the scope of my concerns for this paper.

A similar set of arguments renders environmental concerns irrelevant to the patent debate. When hearing the *Chakrabarty* case, the Supreme Court declined to comment on the environmental dangers that genetic engineering might pose, claiming that such matters were beyond its mandate as well as beyond the mandate of the patent office itself. Without dismissing or endorsing the claims to potential ecological harms—displacement of naturally occurring species, reduction in genetic diversity, and the potential dangers genetically modified crops or other organisms might pose to human beings—the Court simply excused itself from that discussion.[23] The full scope of environmental risks are not fully known, but this is outside the scope of issues addressed by the patent law.

Genetic engineering is becoming more pervasive, and more organisms are being modified than ever before. The patent system plays an economic role in supporting this kind of research. If one is eager to stop the spread of genetic modification for other reasons, targeting the patenting of organisms may make sense as part of a systematic attack on biotechnology, though it seems doubtful that a complete moratorium on patenting organisms would stop the spread of biotechnology.

The most basic problem with current patent office policy is its failure to strike a balance between protecting real innovation and discovery in modified organisms, and recognizing that organisms are not like normal matter. Genes and organisms as found in nature are highly valuable, and it is difficult to set the standards for modification necessary to such natural products to warrant intellectual property rights in the resulting organism. The vast majority of the structures composing an organism and the processes that sustain it are currently beyond our ability to replicate from scratch. Most of biotechnology is built on recombinant DNA techniques, which use natural biochemical processes and organism components to build organisms with new properties. We have been able to tinker with cells by using natural processes to do new things. The nucleases used to cut DNA are derived from bacteria that use them as a defense against invading viruses. In some kinds of genetic transfer, preexisting bacterial plasmids are used to hold the transplanted DNA. They are not made from scratch. Bacteriophages, a form of bacterial virus, are used to transfect organisms with the transplanted DNA. Conjugation is a natural process that can be used to transfer genetic material as in the bacteria Chakrabarty "invented." The manufacture of a transgenic mouse requires skill, but it does not involve making a mouse from scratch; it is a complicated process that harnesses the normal developmental process of the mice.

These observations are not intended to belittle the achievement of recombinant DNA technology, but only to note that it does not now, or in the forseeable future, involve the *de novo* creation of organisms from raw chemical components, only their modification, albeit unnatural or useful modifications. Tinkering with the preexisting structures, cloning, and the transfer of genetic material across boundaries that usually prevent such transfer involve mastering difficult techniques, but should not be confused with the creation of a new organism from wholly nonliving parts of human manufacture.

Mark Sagoff uses a distinction from John Locke's *First Treatise of Government* (1698) to question the appropriateness of granting patents on genes and organisms. He distinguishes between genuine authorship, a form of creation which brings with it intellectual property rights, and parenthood, which is also a form of creation, but one that does not come with the privileges of authorship because the parents merely serves as "the occasions of their being" for their offspring. Sagoff thinks that biotechnologists fit into the latter category and so do not deserve intellectual property rights in the organisms they produce. "Technologists may claim intellectual property rights in organisms, I believe, insofar as they design them, using ideas that are not found in nature but are their own."[24] He thinks that Chakrabarty's bacteria should have been treated in the same way that the bacteria were treated in the *Funk Bros.* case; the patent should have been denied because Chakrabarty's bacteria were products of nature. "To cross strains of plants or microinject genes into embryos is not to invent, design, or create living things." He denies that Chakrabarty did enough creative work to justify the patent protection he received. He says that "the Patent Office, in other words, failed to distinguish manipulating or changing a genome from designing or inventing it." I am not sure that his conclusion follows from his premises. Certainly Chakrabarty did not design a novel bacterium from scratch. He assembled more or less preexisting parts into a new organism. By the standards of the patent office, though, that was enough to justify the protection Chakrabarty received, and it does seem as if it is more in keeping with the general standards for patentability.

Sagoff further develops his position on animal patents by contrasting the work of biotechnologists with that of God. Without endorsing Sagoff's conclusion about patentability or even accepting his explanation for the apparent design we find in organisms, he does seem to have located a source of discomfort for those who defend these patents. Current organisms have the properties they do because of their evolutionary histories. The apparent design has no real author unless natural selection is reified as an active agent in evolutionary processes. This certainly seems to be true, but should not prevent those who significantly modify those natural givens from protecting that innovation through the use of patents, so long as they do not thereby gain control of the natural given with which they started.

Sagoff also argues against patents on biotech products, including organisms, on the grounds that there is no quid pro quo of knowledge for the patent monopoly granted because the inventor cannot fully explain the natural processes that underlie her invention. His concern seems to be that not all biological inventions can be adequately disclosed through description, so the patent holder gets more than she should because no enabling disclosure has been made. His criticism is a good one if directed at plant patents that only have to be disclosed to the extent possible. Particularly if the plant patent is on a natural sport discovered in a cultivated field, that disclosure may be quite brief and not enough to allow anyone to duplicate the "invention." This problem has been resolved to some extent by allowing a deposit of biological materials to count as adequate disclosure.[25] Further, applicants need only disclose how their improvements to the organism were achieved. They do not have to disclose how to make the organism from scratch any more than the inventor of a new farming implement needs to explain how to smelt the iron ore used to make the implement.

Sagoff is troubled by an important concern—how to ensure that the patent recipient does not get too wide a range of protection for a relatively narrow improvement to an organism. Scientists working for Agracetus, a biotech subsidiary of W. R. Grace and Co., were the first to develop a technique for inserting a foreign gene into cotton plants. In 1992 the company was granted a patent on "all genetically engineered cotton products" which gave them the right to exclude others from making any genetic modification of cotton without a license from them. The patent was eventually overturned in 1994 on the grounds that it was based in prior art.

Sungene, a biotech company, has been granted a U.S. patent for a "sunflower variety with very high oleic acid content. The claim allowed was for the characteristic (i.e., high oleic acid) and not just for the genes producing the characteristic. Sungene has notified sunflower breeders that the development of any variety high in oleic acid will be considered an infringement of its patent."[26]

Both of these cases demonstrate that judgments about the scope of claims granted and the kind of patent to be issued—product, product by process, or process—are important in all areas of patent law, but seem to have been particularly troublesome in granting patents on genetically modified organisms. If a patent is written narrowly, it offers little protection; others can invent around it. For example, if there are a dozen known gene sequences that can increase oleic acid production in sunflowers, all of which can be spliced in with ease, then the value of a patent that excludes others from using only one of those sequences is much less than a patent that excluded others from using any of the dozen known sequences. Even better would be to be able to exclude others from producing a sunflower with increased oleic acid content by any method. I do not have a formula that would allow examiners to determine in advance the right balance between protection for real innovation and not blocking the later innovations of others. I would be suspicious of anyone claiming to have such a formula, but it seems that in such a quick-moving field, innovation will be promoted best by erring on the conservative side, granting relatively narrow process patents and few product patents.

The typical applicant for a patent on an organism has built onto or altered an object of an unusual kind. We do not have a complete understanding of biological development and are nowhere near being able to recreate even the simplest of organisms from scratch. In some ways it is equivalent to a case in which an innovation is made in a complicated machine that is not already protected by patent for one reason or another—let's say it fell from space or had been devoted to the public. If someone were to find such a machine and make an improvement to it that is useful, nonobvious, and innovative, and could be effectively disclosed, there is no reason why that person should not be granted a patent on the *improved* machine, but only on the improved machine, leaving the preexisting object in the public domain. Such a patent ought not to prevent other inventors from altering the preexisting machine for which no one is able to claim intellectual property rights. The patent applicant should not get rights to more than he has added to the preexisting organism; why permit the patentee to skim the cream? Patents should be approved only if written in such a way as to allow further innovation using the same nonpatentable source material.

Another concern about organism patents focuses on what seems to be a land grab of previously commonly held resources. It is worth remembering that in a majority of these cases, the source material remains in the public domain and only the modified products are protected. This protection may seem inadequate if you yourself are the source of the product to be modified or a member of a culture that has developed the source organism and do not stand to benefit from the variations that have been developed.

In 1976, for example, after John Moore had a splenectomy as part of his treatment for hairy cell leukemia, he was informed by the California Supreme Court that he did not own cells taken from his body.

> On several visits over the next seven years, at [Dr.] Golde's direction, Moore returned to the [UCLA] Medical Center. During these visits, Golde withdrew samples of Moore's blood, skin, bone marrow, and sperm. Golde told Moore that these samples were required to monitor and ensure Moore's continued health.[27]

Moore's doctor, David W. Golde, established a cell lineage from his spleen and applied for a patent on January 6, 1983, for "a Unique T-Lymphocyte Line and the Products Derived Therefrom." The patent was granted on March 20, 1984.[28] Moore discovered what had happened and sued. The case ended in the California Supreme Court (*Moore* v. *Regents of the University of California*) where it was ruled that Moore did not own his cells or have any financial interest in what happened to them.[29]

Companies in the developed world hold the overwhelming majority of patents on naturally occurring ingredients from the Southern Hemisphere.[30] The active components of several folk remedies and crops well known in India have been patented or attempts have been made to patent them in the United States, for example, the use of turmeric to heal wounds, an herbal concoction with antidiabetic properties, and a new variety of basmati rice. The turmeric patents were eventually overturned when written accounts of its use in folk medicine were discovered in India, but as of this writing the other two still stand. This appropriation of traditional knowledge is made possible by an asymmetry in the U.S. Patent and Trademark Office policy. Although written mention of the proposed invention will prevent a patent, whether that mention is domestic or not, only domestic knowledge or use without written mention can prevent a patent. India recently began a program to document traditional knowledge in a digital database accessible to the U.S. Patent and Trademark Office to establish prior art to block this kind of patent.[31]

These cases involve different sorts of exploitation. In Moore's case, he himself was used as a source of patentable materials. More common are cases like that of turmeric or basmati rice in India or botanical pharmaceuticals from the Amazon basin in which traditional knowledge is acquired abroad and then patented in the United States. Some kind of protection for these sources of patentable materials is warranted. Removing the distinction between domestic and foreign folk knowledge would be a big step in the right direction. As for Moore, it makes sense that he would not himself receive a patent on the cells taken from his body. Using Sagoff's earlier distinction, Moore is clearly the occasion for their being; he is not their author or inventor, but his informed consent should have been required before his doctor could develop or patent the cell lineage derived from him.[32]

I would like to address one more issue, which is the effect of biotechnology patents on basic science. No systematic data are available, but complaints have been widespread that the patent system, particularly patents on genes and organisms, are interfering with the underlying social structures of basic research.[33] The changes in the patent law, and even more so in its interpretation by the courts and patent office, have added legal concerns about intellectual property rights to areas of science that once operated in relative innocence. Workers in basic science have always wanted credit for their work, but that credit had not previously come in the form of a patent.[34] Many biotech inventions skirt the line between basic research and technological applications. Into which category, for instance, does the OncoMouse fit? It was the first recombinant DNA mammal, but it is also a helpful lab tool for studying carcinogens.

Before Congress passed the Bayh–Dole Act (a.k.a. Patent and Trademark Laws Amendment), fewer than 250 patents were issued to universities in the United States each year. It was extremely complicated to transfer patents from the public to the private sector or to get patents for government-sponsored research. Two motivations for the legislation were the perception that federal labs were not efficiently transferring information to those who could use it, and the unclear patent status of federally supported university research. Bayh–Dole allows universities to patent and commercialize products, processes, and the like, that were developed during federally funded research, as long as they meet certain requirements, for example, the government gets a license on the resulting product too. Universities in the United States now receive about 1,500 patents a year that mention Bayh–Dole as part of the special conditions that spawned the biotech revolution.[35] Bayh–Dole requires universities to report any potential patentable material to the sponsoring agency to share any resulting royalties with the inventors.

The Constitution does not make the distinction between use of a patented invention to compete commercially with the patent holder, and its use in pure science. Judicial precedent, however, supports an exemption for some uses. In the 1813 case of *Wittemore* v. *Cutter*, Justice Joseph Story argued that "It could never have been the intention of the legislature to punish a man, who constructed a [patented] machine merely for philosophical experiments, or for the purpose of ascertaining the sufficiency of the machine to produce its described effects."[36]

This informal exemption has received a narrow judicial interpretation. Cases like these rarely make it to court because patent holders tend not to enforce their rights against basic scientific uses.[37] This exemption would not hold if the patented product was one intended for use in basic research.[38] When this experimental use exemption is used in court as a defense, it is almost universally unsuccessful because it seems only to be used by competing commercial interests who have been caught with their hands in the cookie jar.

How should patent law interact with the needs of basic research? The patent holder has no compelling economic interest in patent exemptions for research. But science seems to progress best when there is free access to prior discoveries. So long as there is a clear line between basic research and commercial inventions, this kind of difficulty is not a big deal. But, this always gray distinction has become even more vague due to the increasingly promiscuous distributions of patents for what might once have been thought of as unpatentable basic research. In practice today the "usefulness" requirement of the rules for patentability are essentially unenforced—almost any putative use will suffice, making it easier to patent research before it produces practical applications.

Securing patent protection for basic research and then enforcing those patents against other basic researchers obviously flies in the face of the Mertonian norms of science and can short-circuit the normal reward system of pure science. Merton discusses this explicitly in "The Normative Structure of Science" (1942), and his arguments constitute the base for much legal scholarship on the subject.

> The increasing value of patents makes adherence to the traditional community norm of nonproprietary open access implicitly more expensive. Thus, even if a particular scientist believes strongly in adherence to the norm, he or she knows others will be tempted to ignore it because of the higher payoff that stems from seeking a patent.[39]

It becomes even more difficult to enforce these norms if there is a potentially big payoff to violating them and little means of enforcing them on others. Free access is also supposed to allow for replication and as a test against fraud and error. Scrutiny is most likely when people disagree and have something at stake. Scientists threatened by your results are going to do a better job of checking over your work than your friends.

> But as the line between basic and applied research becomes blurred in certain fields, patent protection increasingly threatens to encroach on the domain of research science, making it necessary to work out an accommodation between the two perspective. A carefully formulated experimental use exemption from patent infringement liability is an important first step in that direction.[40]

The line between basic research that is unpatentable and applied research that is patentable is particularly vague in biotechnology: "Scientists working in biotechnology-related fields are increasingly likely to be concerned simultaneously with the norms and rewards of research science and the rules and incentives created by intellectual property law."[41] A research exemption is currently being considered in Germany and may serve as a good case study for possible changes in the United States.

APPENDIX
TIMELINE LEADING TO AND FROM THE CHAKRABARTY PATENT DECISION

1813. *Wittemore* v. *Cutter*—"It could never have been the intention of the legislature to punish a man, who constructed a [patented] machine merely for philosophical experiments, or for the purpose of ascertaining the sufficiency of the machine to produce its described effects" (29 F Cases 1120 (D Mass 1813).

1817. *Lowell* v. *Lewis*—"All that the law requires is, that the invention should not be frivolous or injurious to the well-being, good policy, or sound morals of society. . . . The word 'useful,' therefore is incorporated into the act in contradistinction to mischievous or immoral. For instance, a new invention to poison people, or promote debauchery, or to facilitate private assassination, is not a patentable invention."

1854. *O'Reilly* v. *Morse*—The Supreme Court held that Samuel Morse could not patent "electromagnetism, however developed for marking or printing intelligible characters, signs, or letters at any distances" 56 US (15 How) 62 113 (1853).

1873. U.S. patent 141,072 granted to Pasteur for "yeast, free from organic germs of disease as an object of manufacture."

1889. *Ex parte Lattimer* (plants, even those artificially bred, are products of nature for the purposes of patent law) decision by the commissioner of patents. The applicant had tried to patent "cellular tissues from the

Pinus Australis" separated from other components of the plant to be spun into fibers. The claim was rejected as being a product of nature and thus not patentable subject matter. Prior to passage of the Plant Protection Act, this decision led to the popular belief that plants, however modified, are products of nature and so not patentable.

1928. Appeals court decides that General Electric cannot patent pure tungsten even if it is not found in nature, in *General Electric, Co.* v. *DeForest Radio, Co.* 17F (2) 90 (D Del).

May 13, 1930. U.S. Congress passes the Plant Protection Act. First patent rights for agriculture in the world. Novel varieties of plants can be patented, giving the patent holder protection against unauthorized asexual reproduction of the patented plant.

March 2, 1931. In *American Fruit Growers* v. *Brogdex Co.* 283 US 1, 11, the Supreme Court rules that an orange treated with a solution of borax is not an article of manufacture for patent law purposes.

1933. Patent 18,941,351 granted on a yeast preparation.

1934. *City of Milwaukee* v. *Activated Sludge*, patent on a septic tank process involving live bacteria.

1940. *In re Arzberger* (112 F 2d 834, 46 USPQ 32 (CCPA 1940), the CCPA rules that bacteria are not plants for the purposes of the PPA and so not patentable under that statute.

1942. Patent 2,271,819 granted on a distemper virus vaccine.

1948. *Funk Brothers Seed Co.* v. *Kalo Inoculant.* The Supreme Court rules that combining bacteria is insufficient to count as an invention.

1950. Patent granted on a hog cholera virus (2,966,433).

1954. Congress amends the PPA to include patents on "newly found seedlings" and to exclude "a plant found in an uncultivated state."

1963. Patent 3,133,066 granted for a composition containing oil and *Bacillus thuringiensis* spores.

1970. Congress passes the Plant Variety Protection Act (PVPA), which extends patent-like protection to sexually propagated plants.

H. O. Smith isolates the first restriction enzyme.

Cetus is founded.

In re Argoudelis decided in the U.S. Court of Customs and Patent Appeals. The decision allows for microorganisms to be disclosed in accord with U.S. Code Title 35 Section 112 when applying for patent protection through depositing cultures of the microorganism in a public depository in the United States.

1972. Chakrabarty applies for patent on an oil-consuming bacterium.

Jackson, Symons, and Berg splice DNA from one virus into another. This is the first time DNA has been joined from different organisms in vitro.

1973. First DNA is spliced into a plasmid and inserted into an *E. coli*

1974. *In re Mancy*, the Court of Customs and Patent Appeals rejects a claim for a newly discovered and isolated (biologically pure) culture of a microorganism, though allowing the patent on a process involving the microorganism. Decision later criticized in *In re Bergy* by the CCPA.

February 1975. Asilomar Conference on Recombinant DNA safety.

1976. Genentech is founded.

Chakrabarty case decided in the patent board of appeals, patent denied. Chakrabarty appeals decision to the CCPA.

John Moore's spleen is removed as part of his treatment for hairy cell leukemia.

March 2, 1978. The Court of Customs and Patent Appeals reverses earlier *Chakrabarty* decision by the board of appeals of the patent and trademark office. The patent office appeals and the case is remanded by the Supreme Court back to the CCPA for further consideration in light of *Parker* v. *Flook*.

March 29, 1979. The Court of Customs and Patent Appeals again upholds Bergy's and Chakrabarty's claims. Patent office appeals decision to the Supreme Court, which agrees to hear the case. The majority opinion finds no relevance in *Parker* v. *Flook*.

1980. Congress passes the Bayh-Dole Act (a.k.a. Patent and Trademark Laws Amendment), making it much easier for university scientists to patent work done while receiving government money.

First transgenic mice produced by direct injection of cloned DNA into the pronucleus of a fertilized egg.

Bergy withdraws his claim and his case is declared moot. Bergy may have withdrawn his claim as a tactic by Upjohn to have Chakrabarty heard alone because it was a stronger case for patentability.

Amgen founded.

Cohen-Boyer patent (4,237,224) granted for a process for producing biologically functional molecular chimeras.

Supreme Court resolves *Diamond* v. *Chakrabarty*. Patent granted.

1982. The Court of Appeals for the Federal Circuit (CAFC) is established.

Foreign gene inserted into a tobacco plant and transmitted to the plant's progeny.

January 6, 1983. Golde applies for a patent on cell line derived from Moore's tissue.

1984. Patent filed on the Harvard OncoMouse.

Patent (04,438,032) granted on the cell line derived from John Moore's spleen.

Moore sues his doctor, the University of California, and two biotech firms.

1985. PTO Board of Patent Appeals and Interferences decides *Ex parte Hibberd*. Nonnaturally occurring man-made multicellular plants are patentable under section 101. Reason stated is the *Chakrabarty* decision.

1986. Congress passes the Technology Transfer Act, which makes it easier for government scientists and agencies to patent their discoveries.

April 3, 1987. *Ex parte Allen* (2USPQ2d 1425 (Bd Pat App & Interf 1987) (polyploid oyster). The patent appeals Board denies a patent for a polyploid oyster on the grounds that it was obvious in light of prior art, but raises no objections to the fact that one of the claims is for a multicellular animal. *Chakrabarty* decision is again cited.

April 7, 1987. Quigg's news release stating that "The Patent and Trademark Office now considers non-naturally occurring non-human multicellular living organisms, including animals, to be patentable subject matter within the scope of 35 U.S.C. 101."

April 12, 1988. U.S. patent 4,736,866 granted on the "OncoMouse." Harvard gets the patent on the genetically altered mouse developed by Leder et al., "transgenic non-human mammal all of whose germ cells and somatic cells contain a recombinant activated oncogene sequence introduced into said mammal, or an ancestor of said mammal, at an embryonic stage."

April 26, 1989. Europe allows the first plant patent to Lubrizol Corp. for a modification to the genes of sunflowers, alfalfa, and soybeans that makes them store more protein.

July 1989. European Patent Office rejects patent on the "OncoMouse."

July 1990. *Moore* v. *Regents, University of California*, California Supreme Court rules that Moore has no property rights in the cells taken from him.

1991. European Patent Office (EPO) grants patent on the Harvard Onco-Mouse.

NIH applies for patents on over 2,500 DNA segments that it sequenced in order to protect them for development that would be difficult if they were committed to the public domain.

1992. Agracetus, a biotech company, is granted a patent (5,159,135) on any genetically modified cotton plant or seed.

Craig Ventor forms Human Genome Sciences.

Patent granted on a transgenic chicken resistant to avian leukosis virus.

February 1993. NIH applies for patent protection for 2,421 genetic sequences representing parts of genes, most of them without knowing the functions of those genes.

Patent 5,183,949 granted on a rabbit model for HIV testing.

1994. Agracetus patents canceled by U.S. patent office, NIH abandons its patent application for the genetic sequences.

1995. Congress passes the Biotechnological Process Patent Act to allow certain biotechnological process claims without consideration of obviousness, where the process uses or results in a composition of matter that is both novel and nonobvious.

November 13, 1995. *Imazio Nurseries, Inc.* v. *Dania Greenhouses* settles the extent of PPA protection to those plants produced by asexual reproduction from the original plant based on differing definitions of "variety" in the PPA and PVPA.

February 1997. The NIH abandons its patent on a human cell line derived from Papua, New Guinea, placing it in the public domain.

December 18, 1997. Stuart Newman and Jeremy Rifkin file a patent for creating human-animal chimeras and implanting them into surrogate mothers.

March 3, 1998. Patent granted on the "terminator" gene.

April 21, 1998. Patent 5,741,957 granted on a transgenic cow.

October 20, 1998. Patent 5,824,841 granted on a tetraploid oyster.

November 1998. Canada decides that the OncoMouse in unpatentable. Among the reasons stated are the extent to which the mouse represents more natural process than human intervention, which is limited to the microinjection.

December 15, 1998. Patent granted on a rat with straight filaments in its brain.

March 10, 1999. Federal Appeals Court agrees to hear *Pioneer Hi-Bred* v. *Farm Advantage* in a case set to test the validity of plant patents and the precedent of *Ex parte Hibberd*.

April 15, 1999. Consortium of ten of the world's largest drug companies and five leading gene laboratories agree to exploring common human genetic variation and placing the information into the public domain. Altruism? Maybe, but it also starves and preempts small biotech companies.

June 1999. *Regents, University of California* v. *Genentech, Inc.* Patent infringement case related to human growth hormone and the theft of material from the University of California-San Francisco ends with an eight-to-one deadlocked verdict against Genetech.

Rifkin and Newman's patent application rejected by PTO in part because "the claimed invention embraces a human being." Rifkin refiles the patent as a step towards an appeal of the decision.

NOTES

1. Edmund Kitch, "The Nature and Function of the Patent System," *Journal of Law and Economics* 20 (1977):265, said that the additional benefit of the system is to give investors prospect rights to work a claim for a number of years. The basic idea is that the scope of the patent granted should protect, and therefore encourage, future research by the patent holder into areas closely related to the protected invention while not protecting a greater stake than the patent holder could be expected to explore effectively.

2. Robert G. Bone, "A New Look at Trade Secret Law: Doctrine in Search of Justification," *California Law Review* 86 (March 1998), provides a good summary of U.S. trade secret law (pp. 241–310). "To qualify as a trade secret, information must meet three requirements: (1) it must confer a competitive advantage when kept secret; (2) it must be secret in fact; and (3) in many states, it must be protected by reasonable secrecy safeguards" (p. 248). Trade secrets are protected only from misapproporation, not from reverse engineering or independent discovery.

3. Paul Rabinow, *Making PCR: A Story of Biotechnology* (Chicago: University of Chicago Press, 1996), 25.

4. A complete history of this matter cannot be written without desribing the role of another patent application, that of Malcolm Bergy for a biologically pure culture of the microorganism *Streptomyces vellosus*. Bergy had discovered this organism and created the first biologically pure culture of it.

5. "In the Matter of the Application of Ananda M. Chakrabarty," Appeal No. 77-535, U.S. Court of Customs and Patent Appeals, 2 March 1978.

6. Meanwhile, Bergy had applied for a patent on a microorganism he discovered and bred as a pure culture. His microorganism was deemed a product of nature and the examiner denied his application. The board of appeals of the U.S. Patent and Trademark Office affirmed the decision but only on the grounds that living organisms were not patentable under Section 101. The CCPA reversed the board because the biologically pure culture was not found in nature and was deemed a manufacture or composition of matter as defined by Section 101. The CCPA's decision was not unanimous, just as in *Chakrabarty*. The government petitioned for certiorari in *Bergy* and got it in 1978. The case was remanded for further consideration in light of *Parker* v. *Flook*. Because they were similar, the CCPA vacated the

Chakrabarty decision and heard the two cases together. The CCPA upheld the claims of patentability by *Chakrabarty* and *Bergy*. Diamond, the commissioner of patents, requested certiorari again and was granted it for both cases. Bergy withdrew his claim and the case was dismissed as moot (summary from Miggs, 1981).

7. *Funk Brothers Seed Co.* v. *Kalo Inoculant*, 33 U.S., 131

8. The Supreme Court, like the CCPA, did not emphasize or even take into consideration the usefulness of Chakrabarty's bacteria. The role of patents in scientific progress and the fact that the patent system as delineated by the Constitution is intended to promote science were not discussed.

9. The Plant Protection Act was originally known as the Townsend-Purnell Plant Patent Act, passed after plant breeders lobbied Congress to protect their work. See the complaint of David Fairchild, president of the American Genetic Association in the February 1927 issue of *Heredity* and letters from New York plant breeders and nurserymen appended to the Senate Proceedings for May 12, 1930, pp. 8750–51. I'm not sure I can fully substantiate this claim, but the fact that patents were now being granted on chemical compounds may have been a part of the motivation too. See the congressional record cited in *Chisum on Patents* 105 [1], 1–250.

10. Letter from Burbank's widow passed on to Paul Stark. Burbank was not entirely consistent on this issue. "No patents can be obtained on any improvements of plants, and I for one am glad that is so. The reward is in the joy of having done good work, and the impotent envy and jealousy of those who know nothing of the labor and sacrifices necessary, and who are by nature and cultivation, kickers rather than lifters." *How to Judge Novelties*, issued January 1911, quoted in Peter Dreyer's *A Gardener Touched With Genius: The Life of Luther Burbank* (New York: Coward, McCann, & Geoghegan, 1975).

11. Standards for "new variety" include resistance to disease, drought, cold, or heat, color, flavor, perfume, or productivity. Not all of these are easily disclosed in writing or a line drawing; this presented difficulty in "disclosing" the new plant.

12. When the PPA was passed, it was viewed as a radical piece of legislation. A number of patents were retroactively granted to Luther Burbank, who had died several years earlier, on plants he developed prior to the enactment of the PPA. And Harold Ickes Sr., secretary of the interior, was granted a patent on a hybrid dahlia. Senator Dill quoted from the Congressional Proceedings for April 14, 1930, p. 7017: "May I say that my reason for speaking was simply that the Senate might understand the remarkable kind of legislation it is. I do not want alone to take the responsibility of stopping the passage of the bill, for it may be that my doubts are not justified; but I have felt that it is such a departure from anything we have ever done in the Senate that the Senators ought to realize what kind of legislation it is." And in the House Proceedings, May 5, 1930, p. 8391–92, LaGuardia said, "The gentleman (Purnell) will concede that this is rather novel, will he not?" and "I think we should study this a great deal."

13. Robert Starr Allyn, *The First Plant Patents: A Discussion of the New Law and Patent Office Practice* (Brooklyn: Educational Foundations, 1934), 28.

14. Robert C. Cook, "The First Plant Patent," *Journal of Heredity* (October 1931): 313–16.

15. The act also encouraged breeders to asexually reproduce their specimens so that they might be patented.

16. This act was passed in response to the 1961 International Union for Protection of New Varieties of Plants. For exceptions to the PVPA (carrots, celery, cucumbers, okra, peppers, tomatoes—all repealed in 1980), see Lawrence Busch et al., *Plants, Profit, and Power: Social, Economic, and Ethical Consequences of the New Biotechnologies* (Oxford: Blackwell, 1992), 27.

17. 2 US PQ 2d at 1426-1427. (Quoted from Chisum 1-51).

18. An oyster with three sets of chromosomes has since been patented.

19. The status of partially human chimeras has not yet been established. Jeremy Rifkin and Stewart Newman have applied for a broad patent on methods for creating human-animal chimeras. The patent application was rejected and is currently being appealed.

20. Kenneth J. Burchfiel, *Biotechnology and the Federal Circuit* (Washington, D.C.: BNA Books, 1995), 34.

21. There is the judicial precedent of *Lowell* v. *Lewis* (1817) to support moral opposition to patents. "All that the law requires is, that the invention should not be frivolous or injurious to the well-being, good policy, or sound morals of society. . . . The word 'useful,' therefore is incorporated into the act in contradistinction to mischievous or immoral. For instance, a new invention to poison people, or promote debauchery, or to facilitate private assassination, is not a patentable invention." This is a dubious interpretation of the patent statutes. Useful is in contradistinction with useless, not immoral. Usefulness is rarely imposed as a substantial standard. See Chisum 1997 (4.03). Usefulness can be determined by market forces if other conditions for patentability are met.

22. Mark Sagoff, "Animals as Inventions: Biotechnology and Intellectual Property Rights," *Institute for Philosophy and Public Policy* 16, no. 1 (winter 1996).

23. Ibid.

24. In 1970, *in re Argoudelis* was decided in the U.S. Court of Customs and Patent Appeals, allowing for inventions involving microorganisms to be disclosed in accord with U.S. Code Title 35 Section 112 when applying for patent protection through depositing cultures of the microorganism in a public depository in the United States.

25. Vandana Shiva, *Biopiracy: The Plunder of Nature and Knowledge* (Boston: South End, 1997), 55.

26. E. Richard Gold, *Body Parts: Property Rights and the Ownership of Human Biological Materials* (Washington, D.C.: Georgetown University Press, 1996), 23.

27. Moore was asked to sign a consent form: "I (do, do not) voluntarily grant to the University of California any and all rights I, or my heirs, may have in any cell line or any other potential product which might be developed from the blood and/or bone marrow obtained from me."

28. See Gold, *Body Parts.*

29. Seth Shulman, *Owning the Future* (Boston: Houghton Mifflin, 1999), 133.

30. See K. S. Jayaraman ". . . and India Protects Its Past Online" in *Nature* 401 (30 September 1999): 413–14.

31. See also Karen Gottlieb, "Human Biological Samples and the Laws of Property: The Trust as a Model for Biological Repositories" in *Stored Tissue Samples: Ethical, Legal, and Public Policy Implications,* ed. by Robert F. Weir (Iowa City, Iowa: University of Iowa Press, 1999).

32. "Indeed, the competition has even intensified, with the recognition that there are useful things to do with molecular biology, and money to be made as well. None of us can guess what the benefits will be, except that they will be huge. The old innocence has long since gone. Will academic's corporate interests now undermine the integrity of the research university, especially at a time when governments everywhere are ready to shuffle responsibility to any other economic sector willing to pay the bills?" John Maddux, editor emeritus, *Nature* in the preface to the expanded edition of Horace Judson's *The Eighth Day of Creation* (Woodbury, N.Y.: Cold Springs Harbor Laboratory, 1996).

33. David Hull describes the social structure behind basic science in *Science as a Process: An Evolutionary Account of the Social and Conceptual Development of Science* (Chicago: University of Chicago Press, 1990).

34. Scientists exchange credit for contributions. Credit comes in a variety of forms from prestigious prizes to citations. One sort of credit is most fundamental—the use that one scientist makes of the work of another. The "success" that is central to science is not career advancement but mutual use. Science has the cumulative character that it has in part because of this sort of credit.

35. Christopher E. Bush, "Medical Patent Battle Leads to Bayh-Dole Act Test Case; Dispute Hinges on Innovative, Lucrative Cancer Treatment," *Corporate Legal Times,* January 1998, 19; Rabinow, *Making PCR.*

36. When an alien system that rewards secrecy until patenting is added to this set of incentives, the results can be upsetting to the established social structure of science; *Wittemore v. Cutter* 29F Cases 1120 (D Mass 1813).

37. Rebecca Eisenberg, "Patents and the Progress of Science: Exclusive Rights and Experimental Use," *University of Chicago Law Review* 56 (1989): 1017–86.

38. See also *Roche* v. *Bolar Pharmaceutical Co.* in which the distinction is made between "unlicensed experiments conducted with a view to the adaption of the patented invention to the experimenter's business" as opposed to experiments conducted "for amusement, to satisfy idle curiosity, or for strictly philosophical inquiry."

39. See also Robert Merge, "Property Rights Theory and the Commons: The Case of Scientific Research" from *Scientific Innovation, Philosophy, and Public Policy* (New York: Cambridge University Press, 1996), "Surely it would be stretching quite a bit to argue that the presence of patents does not make a difference in the conduct of science. Yet just as surely it would be wrong to say that patents lead researchers to completely shut off the exchange of research results. Nor are patents universally enforced to the hilt among researchers; far from it." (p. 150)

40. Eisenberg, "Patents and the Progress of Science."

41. Rebecca Eisenberg, "Propietary Rights and the Norms of Science in Biotechnology Research," *Yale Law Review* 97 (1987): 177–231.

three

ETHICAL ISSUES AND APPLICATION OF PATENT LAWS IN BIOTECHNOLOGY
ROCHELLE K. SEIDE
AND CARMELLA L. STEPHENS

INTRODUCTION

Since the advent of genetic engineering in the 1970s, major break-throughs have been made in the fields of genetics and cell biology. Newly developed cloning techniques have provided scientists with opportunities to genetically engineer altered life-forms, resulting in an abundance of practical applications in the fields of medicine and agriculture. For example, new treatments for a wide range of different diseases, improvements in crop yields, and genetic testing for predisposition to genetic disease have been made possible due to progress in the field of genetic engineering. Recombinant erythropoietin, a protein capable of stimulating red blood cell production, is now routinely used to treat patients with kidney diseases and anemia resulting from cancer chemotherapy. Similarly, genetically engineered insulin has replaced insulin that had previously been purified from bovine and porcine sources. The discovery of genes predisposing people to diseases including cancer of the breast and colon has generated new diagnostic tests. Genetically engineered cattle and goats produce therapeutic proteins in their milk, and farmers are planting genetically engineered crops that are resistant to plant disease and insect infestation, thereby reducing the need for use of environmentally toxic pesticides and insecticides.

The successful development of new therapeutics, diagnostics, and crops has helped to fuel biotechnology, resulting in the creation of a major industry. Of significant importance in the development of this industry has been the ability to obtain patent protection on biotechnology-related innovations. Such patent protection is essential as it reduces commercial risks and provides economic incentives to commitment of capital and human resources.[1] For the biotechnology industry, this is particularly important given the great expenditure of time and money required to develop a new drug. Patents issued to date include those covering genetically engineered therapeutic proteins such as insulin, erythropoietin, tissue plasminogen activator, and growth hormone; transgenic organisms such as the "Onco-Mouse"; disease resistant transgenic plants; and diagnostic methods based on the identification of disease associated genes and methods of gene therapy.

The commercialization of different aspects of life has resulted in a great deal of controversy as to whether patent protection should extend to such subject matter. There are concerns that human life may be losing its sanctity and becoming merely a commercial commodity.[2] This controversy has been fueled in recent years by a number of significant advances in the biological sciences that have raised questions concerning "what is life?"

Such advances include the reported cloning of the sheep referred to as "Dolly."[3] The cloning technique utilized for the creation of Dolly involved the transfer of a nucleus derived from a differentiated somatic cell of an adult animal into an unfertilized donor egg from which the nucleus had been removed. Offspring resulting from this technique are identical genetic twins of the adult animal from which the nucleus was derived. Thus, the process allows for the production of an unlimited number of precise duplicate organisms.

The successful cloning of Dolly raised ethical concerns of possible application of the cloning technology to the cloning of humans. Such cloning is principally objectionable because it creates persons for the instrumental benefit of others thereby violating widely accepted moral principles about the intrinsic value of human beings.[4]

Additionally, significant debate has developed in recent years from research involving the derivation or use of human embryonic stem (ES) cells and embryonic germ (EG) cells.[5] The value of stem cells is attributed to their ability to produce almost any desired tissue when treated with the appropriate growth factors. A single human embryonic stem cell "has the potential—the genetic blueprint and the biological know-how—to become any cell, tissue, or organ in the human body."[6] Given the potential ability of stem cells to generate new human tissues and perhaps entire organs, it is no wonder that *Science* magazine designated stem-cell technology the "breakthrough of the year."[7] It is noteworthy that two patents have recently issued to the University of Wisconsin covering purified primate embryonic stem cells and methods for isolating such cells (U.S. Patent Nos. 6,200,806 and 5,843,780).

Much of the controversy surrounding the use of stem cells revolves around the methods used for deriving such cells. For example, ES cells are derived from early human embryos (such as those created for in vitro fertilization) and EG cells from aborted fetuses. The fear is that embryos will be produced merely for their commercial value or as scientific research tools. To those who believe that life begins at conception, this result is clearly objectionable.

Further, as the complete genomic sequences from a variety of different organisms become available, it will be possible to determine the minimal set of essential cellular genes required for basic metabolism and self-replication. Such experiments "attempt to point not just to DNA generally as the stuff that determines an organism's characteristics, but to a specific set of sequences that defines the difference between life and non-life."[8] Determining and creating a minimal genome would represent the first step in genetic engineering that could ultimately lead to the creation of new organisms. Such experiments seem even closer at hand given that the minimal number of genes required for the survival of a mycoplasma (a parasitic bacteria) has recently been reported in *Science*.[9]

Finally, as is well known, it is the goal of the Human Genome Project to map the entire human genome. As whole or partial DNA sequences are derived, patent applications are filed to protect the entire nucleotide sequence of protein coding DNA. Patent applications often include expressed sequence tags (ESTs) and various polymorphisms.[10] The commercial value of any patents issuing from these applications stems from their importance in detection and screening of disease genes or their role in how people respond to drugs.

One of the controversies surrounding the issuing of patents covering naturally occurring mutations that correlate with an increased risk of developing disease is the belief that they will be used to prevent doctors from practicing medicine. Some fear that patents on fragments of DNA might also impede basic research due to complicated licensing webs. It is also argued that the human genome, as well as naturally occurring mutations within the genome, are inherent to each individual and therefore should not be patentable subject matter. The belief is that "knowledge that a particular DNA sequence at a specific locus is associated with a disease, regardless of the level of effort needed to develop that knowledge, is non-patentable, because it is merely an observation of a state of nature or nature's handiwork."[11]

This seems to illustrate the common misconception that biotechnology patents are attempts to protect "discoveries" rather than inventions. This misconception was recently addressed by Judge Randall Rader of the United States Court of Appeals for the Federal Circuit at a recent conference at the National Academy of Sciences in Washington, D.C. As asserted by Judge Rader, there is no validity to the suggestion that developments in biotechnology are unpatentable "discoveries" rather than "inventions." Further, a key legal distinction to be made is that biotechnological inventions are new entities. Patentable genetic material had never existed in nature in its inventive form until it was "removed, isolated, purified and understood by man."[12]

In addition, it is noteworthy that genetic material, such as a DNA molecule, is really no different from any other naturally occurring chemical entity, such as an antibiotic, hormone molecule, or vitamin, each of which has been found to fall within the statutory categories of patentable compositions of matter. To be patentable, however, the DNA molecule must be in a nonnaturally occurring state resulting from human intervention. Thus, a gene in its isolated form is patentable by virtue of its nonnaturally occurring state and its acquiring of novel characteristics that distinguishes it from its counterpart in nature. Such characteristics include, for example, use as a disease marker or use for recombinant expression of a therapeutic protein.

Recent advances in genetic engineering and biotechnology, such as the cloning of Dolly, the derivation and use of embryonic stem cells, and the sequencing of the human genome, have raised ethical concerns concerning whether this type of research should continue. Those opposed to such research have suggested that the patenting of inventions resulting from genetic engineering should be prohibited. In reality, the concerns of many of those opposed to biotechnology have more to do with the advances being made in biomedical research and the use of such technology, rather than with patent law.

As will be discussed in detail below, the courts over the previous thirty years have addressed many of the issues concerning the scope of patent protection to be extended to inventions resulting from genetic engineering. The courts have concluded that when deciding whether to issue a patent, the role of the U.S. Patent and Trademark Office (PTO) is not to consider ethical and moral issues, but whether the invention fulfills the statutory requirements for patentability. Thus, rather than pressuring the PTO to consider ethical and moral issues, opponents of biotechnology should focus on lobbying Congress to pass legislation that regulates the types of research performed by scientists.

LEGAL BASIS FOR PATENTS IN BIOTECHNOLOGY

Pursuant to the United States Constitution, the PTO is authorized to issue patents "to promote the progress of science and the useful arts by securing for limited time to . . . inventors the exclusive right to their . . . discoveries."[13] A patent for an invention is a grant of a limited property right "to exclude others from making, using, offering for sale, or selling the invention throughout the United States or importing the invention into the United States."[14]

The United States Patent Act authorizes patent protection for any "new and useful process, machine, manufacture or composition of matter, or any new and useful improvement.[15] To secure patent rights to an invention, certain statutory requirements must be met. The standards of patentability necessary to be eligible for patent protection are the same regardless of the field of the invention. The requirements include that the invention be novel, useful, and nonobvious in view of what is known in the field.[16] Additionally, the patent must be described in sufficient detail to enable one skilled in the field of the invention to make and use the claimed invention without undue experimentation.[17]

Currently there are few restrictions on the patenting of different life-forms in the United States. However, initial attempts to patent genetically engineered organisms in the 1970s failed. The PTO rejected the applications on the basis that such organisms were products of nature and, therefore, not patentable subject matter. However, in a 1980 landmark decision, *Diamond* v. *Chakrabarty*, the U.S. Supreme Court ruled that a genetically engineered microorganism was patentable subject matter under U.S. Code Title 35 Section 101.[18] The subject matter of the Chakrabarty patent application was a novel bacterium which was generated using classical microbial genetic manipulation (not recombinant DNA technology) and capable of breaking down multiple components of crude oil. While naturally occurring organisms are not considered patentable subject matter, the

Court interpreted this as not encompassing organisms that have been manipulated by humans. Because the genetically manipulated bacterium possessed the ability to break down crude oil, a property possessed by no naturally occurring bacteria, the bacterium was deemed patentable subject matter. As clearly stated by the Court in *Chakarabarty*, "Congress intended statutory subject matter to include anything under the sun that is made by man."[19] However, as indicated by the Court, this is not to suggest that Section 101 has no limits or it enhances every discovery. "The laws of nature, physical phenomena, and abstract ideas have been held not patentable."[20]

In *Ex Parte Hibberd*, a 1985 case, the U.S. Patent and Trademark Office Board of Patent Appeals and Interferences held that sexually reproducing, man-made plants were patentable subject matter.[21] Thus, the PTO extended utility patent coverage to plants, holding that Section 101 covered seeds, plants, and plant tissue cultures. In a recent challenge to the PTO's ability to grant such patents under the patent statute, the Court of Appeals for the Federal Circuit recently upheld this decision, ruling that patentable subject matter under U.S. Code Title 35 Section 101 includes seeds and seed-grown plants.[22]

While *Chakrabarty* clearly held that genetically engineered microorganisms were statutory subject matter, the patentablility of genetically engineered multicellular organisms had not been tested. In 1987, the PTO Board of Patent Appeals and Interferences decided *Ex Parte Allen*, finding that nonnaturally occurring polyploid oysters in which more than one chromosome set had been induced were statutory subject matter under U.S. Code Title 35 Section 101. Following the *Allen* decision, the PTO specifically stated that it considered nonnaturally occurring, nonhuman multicellular organisms, including animals, to be patentable subject matter. However the PTO added that, "a claim directed to or including within its scope a human being will not be considered to be patentable subject matter under 35 U.S.C. §101."[23] According to the commissioner of patents and trademarks, "The Patent Office made a novel assessment under the Thirteenth Amendment that the grant of a limited, but exclusive property right in a human being is prohibited by the Constitution."[24]

One year later a patent was issued covering the first transgenic animal referred to as "the Harvard OncoMouse." The OncoMouse was developed to be more susceptible to developing tumors, thereby providing a valuable research tool for identifying potentially useful drugs that might be useful for the treatment of cancer. Since then, numerous patents have been issued covering various different types of transgenic animals including mice, pigs, cows, and fish.[25]

Issues concerning the patentability of cell lines and ownership of such cell lines were also addressed by the California Supreme Court in *Moore* v. *Regents of the University of California*.[26] In *Moore*, a patient sued his physician and a biotechnology company for using his biopsied tissue without his consent. Cells derived from the patient's biopsied tissue were transformed into a patented cell line. In *Moore*, the court sided with the defendants, indicating that giving the patient a property right to his tissue would impede progress and destroy the economic incentive to conduct important medical research. The court, however, held that "a physician seeking a patient's consent for a medical procedure must, in order to satisfy his fiduciary duty and to obtain informed consent, disclose personal interests unrelated to the patient's health, whether research or economic, that may affect obtaining his consent."[27]

Whether genes which exist in nature can be patented was addressed by the U.S. Court of Appeals for the Federal Circuit in *Amgen Inc* v. *Chugai Pharmaceutical Co*. In that case, the inventor discovered a method for purification and isolation of DNA encoding erythropoietin, a protein capable of stimulating the production of red blood cells.[28] The test articulated in *Amgen* is that the natural substances are patentable if they are novel, purified, and isolated. As indicated by the court, patent law distinguishes between mere discovery of something existing independently in nature (which is unpatentable) and a true invention (where there is a significant element of human intervention).

In response to concerns about the patenting of nucleic acid molecules, and the scope of such claims, the U.S. Court of Appeals for the Federal Circuit addressed the issue of written description requirements for DNA.[29] In *Regents of the University of California* v. *Eli Lilly and Co.*, the court clarified how the written description requirement of U.S. Code Title 35 Section 112 applies to claims to DNA.[30] In *Lilly* the contested patent disclosed a rat cDNA encoding insulin, but claimed a human insulin encoding cDNA. Although the patent did not disclose the human insulin encoding cDNA, the patent disclosed the known amino acid sequence of human insulin as well as how one would make and use a DNA encoding human insulin. Although the specification satisfied the requirement for teaching how to make and use an invention, the patent was deemed invalid for failing to provide a sufficient

written description of the claimed invention. The *Lilly* court indicated that
a DNA invention is not sufficiently described by merely reciting what the
biological function of a gene is. The inventor must describe the structure of
the DNA by defining its physical or chemical properties.

Due to the efforts of the Human Genome Project to sequence the human
genome, a vast amount of information has been generated including the
sequence of full DNA molecules (genes), expressed sequence tags (ESTs), and
single nucleotide polymorphisms (SNPs). The large number of patent applications that have been filed covering these DNA sequences has raised concerns
about the scope of patent protection that should be given to them. In response
to these concerns and to further clarify the *written description* and *utility* requirements with regard to DNA patents, the U.S. Patent and Trademark Office
recently issued a revised set of examination guidelines.[31]

Under the new guidelines, examiners are to determine whether the
written description requirement has been met. This entails that a skilled
artisan would understand the inventor to be in possession of her claimed
invention, even if every nuance of the claim is not specifically described
in the specifications. Evidence of possession includes the level of skill and
knowledge in the art, whether a complete or partial sequence is disclosed,
physical and chemical properties, functional characteristics, correlations
between structure and function, and method of making.

With regard to the examination of ESTs, the utility guidelines provide
that the scope of patent protection for an EST-type claim will depend
upon the nature and extent of the disclosure. For example, where only a
partial sequence is disclosed, the scope of protection will be narrow and
limited to a disclosed use and the sequences thought to be useful for that
particular use. As the scope of the disclosure increases to include information regarding protein coding regions and activities (i.e., function), the
scope of protection will increase. In addition, under the new utility guidelines, examiners will be required to look for "specific, substantial, and credible utility" in inventions. The new utility guidelines require that (1) the
claimed invention have a specific utility for the claimed invention; (2) that
the utility be credible, and (3) that the utility be substantial. Insubstantial
or nonspecific utilities (referred to as "landfill") do not meet the requirements of U.S. Code Title 35 Section 101 even if such utility is asserted in
the specification. With regard to ESTs, an allegation of the utility of an
EST as a probe without further disclosure is insufficient to meet the utility

and enablement requirement. A disclosure of a specific use that is particular to the sequences contained within the EST is required.

To stimulate attention to questions concerning the ethics of genetic engineering and patenting of engineered life-forms, a patent application was filed in December 1997 claiming a technique for combining human and animal embryo cells to produce a single animal-human embryo. The embryo would be transplanted into a surrogate mother and presumably develop into a chimeric organism. Dr. Stuart Newman and Jeremy Rifkin, a longtime opponent of the biotechnology industry, filed the patent application with the United States Patent and Trademark Office (PTO) claiming the production of a human-animal chimeric that could be up to 50 percent human.

The PTO rejected the patent application for failing to recite statutory subject matter under Section 101 of the patent act, for failing to make enabling and best mode disclosures under Section 112, and for anticipation under Section 102(b). With regard to statutory subject matter rejection, the examiner asserted that

> the claimed invention is not considered to be patentable subject matter under 35 U.S.C. §101 because the broadest reasonable interpretation of the claimed inventions as a whole embrace a human being.[32]

The rejection by the examiner is consistent with the position taken by the PTO that human beings are not patentable.[33]

Following the filing of the chimera patent application, the PTO also issued an advisory indicating that the patent laws would not discriminate against any particular technology, but that there existed strict patentability requirements including utility, novelty, nonobviousness, and adequate disclosure. Citing very early case law, the PTO indicated that with regard to the utility requirement, case law was said to exclude inventions "injurious to the well-being, good policy, or good morals of society." The PTO added "that inventions directed to human/non-human chimeras could, under certain circumstances, not be patentable because, among other things, they would fail to meet the public policy and morality aspects of the utility requirement."

It is worth noting that initially the doctrine of social utility was used to deny patents on gambling devices and products or processes useful only for perpetrating fraud. However, over the years the courts have consistently lowered the utility threshold for an invention with regard to social utility and morality. It is, therefore, of potential concern that the PTO would raise issues of public policy and morality in terms of the utility requirement to exclude an invention from patentablility.

Such a policy would reflect the situation that now exists in Europe where patents on genetically engineered plant and animal species are generally permitted, but patents for any biotechnological applications that "would be contrary to ordre public or morality are prohibited."[34] As set forth in Article 53(a) of the European Patent Convention, a patent will not be granted on "inventions the publication or exploitation of which would be contrary to 'ordre public' or morality." In Europe the requirement that a patent not be granted on inventions that would be "contrary to ordre public or morality" has often been pleaded by opponents of the biotechnology industry, therefore, it is worth briefly examining the history of the morality aspect of the utility requirement in United States patent law.

The requirement that an invention have a "moral" utility was set forth initially by the United States Supreme Court, in *Bedford* v. *Hunt*, where Justice Joseph Story indicated that "by useful invention, in the statute, is meant such a one as may be applied to see beneficial use in society, in contradistinction to an invention, which is injurious to the morals, the health or the good of society."[35] In early cases, an invention lacked morality if it could be used for at least one immoral purpose, for example, as a cover for slot machines.[36] In later cases the test for patentablility was broadened to include inventions that had at least one beneficial purpose.[37]

In *Brenner* v. *Manson*, the Supreme Court lowered the utility threshold still further requiring only that an inventor prove any existing practical use before the granting of a patent.[38] In 1980 the Eighth Circuit Court of Appeals held that even a slight degree of utility is sufficient for patentability.[39] Finally, in *Whistler Corp* v. *Autotronics Inc.*, the court upheld a patent covering a radar detector even given that its sole purpose was to circumvent attempts to enforce speed limits. The court stated that "the matter is one for the legislatures of the states, or Congress, to decide. . . . Unless and until detectors are banned outright, or Congress acts to withdraw patent protection for them, radar detector patentees are entitled to the protection of patent laws."[40]

As clearly indicated, the courts have consistently lowered the utility threshold for an invention. Therefore it is surprising that the PTO would issue an advisory raising issues of morality in the context of utility requirements. The problem with requiring that an invention have a "moral" utility is that it requires subjective analysis that patent examiners identify and apply to patent applications. In addition, issues of morality are constantly changing with time. As expressed by Dr. Robert Merges, "Any one whose life has spanned a decade or two in the twentieth century has witnessed how moral standards can change in a period of a few years. Gambling devices, frowned upon early in the century, are legalized in several states; race tracks and lotteries are now used to generate substantial amounts of income in many states. Birth control devices, in a period of thirty to forty years, have come from a position of illegality to a position where they are welcomed by some as a means of curbing a population explosion."[41]

Although the Rifkin/Newman patent application covering human chimeras has been rejected by the PTO, the filing of the patent application has raised issues which the courts will be forced to address in the near future. For example, although the Thirteenth Amendment would prevent patenting of a human chimera, this would not apply to claims covering the process of producing a chimera. Additionally, the PTO presently grants patents covering human tissues and transgenic animals containing human genes. At what point does a transgenic animal or chimera become human? What percent of the genome must be human derived to qualify as human under the Thirteenth Amendment? These questions may eventually require that the courts define what it means to be human.

It is noteworthy that two patents have recently been issued in Great Britain covering (1) methods of reconstructing an animal embryo via transfer of the nucleus of a donor cell into a recipient cell and (2) the animal embryos prepared using such a method.[42] Surprisingly, the issued claims of the British patents are clearly of sufficient breadth to encompass a human embryo. The European Patent Office also recently issued a patent to "a method of preparing a transgenic animal."[43] In this instance, the patent claim did not include the term "nonhuman" and therefore did not exclude genetic manipulation of humans. In response to protests from European governments and environmentalists, the European Patent Office has stated that the patent mistakenly omitted the term "nonhuman" and as such would technically violate European Union guidelines banning "processes that would change the genetic identity of human organisms." This patent grant will no doubt be challenged in an opposition proceeding before the European Patent Office. Additionally, the PTO recently issued a patent (No. 6,211,429) that covers a method for turning unfertilized eggs into embryos, resulting in the production of cloned mammals. Significantly, the scope of the patent does not exlude humans.

CONCLUSION

Many biotechnological advances to date have satisfied the standards for patent protection as set forth in case law. But as research in the biomedical sciences results in new developments, such as the cloning of Dolly, the derivation and use of embryonic stem cells, and the sequencing of the human genome, ethical considerations and questions concerning the patentablility of such inventions will persist. But as previously stated, moral and ethical issues are not for debate in the area of patent law, but rather are issues that Congress should address while discussing appropriate regulation for biotechnology research. If biotechnology is to be regulated, it should be regulated by Congress, rather than through changes in interpretation of the patent laws by the U.S. PTO.

In fact, there are examples where Congress has sought to regulate biomedical research in response to the ethical concerns of the public. For example, although the PTO has clearly stated that purified and isolated stem cell products and research tools meet the requirement for patentable subject matter, Congress has responded to the controversy raised by the use of embryonic stem cells through legislation. In 1998 a bill was passed by Congress that prohibited the use of U.S. funds for the "creation of a human embryo or embryos for research purposes; or research in which a human embryo or embryos are destroyed, discarded or knowingly subjected to risk of injury or death."[44]

The National Institutes of Health (NIH) recently published draft Stem Cell Guidelines for federally funded researchers that limit research to stem cells derived from excess embryos created for infertility treatments. Areas of research ineligible for funding according to the guidelines include the derivation of pluripotent stem cells from early human embryos, using human pluripotent stem cells to create or contribute to a human embryo, combining human pluripotent stem cells with an animal embryo, transferring a human somatic cell nucleus into a human or animal egg, and using human pluripotent stem cells created for research purposes only.[45] Recognizing the importance of public involvement in the development of policy, the NIH has sought comments on the proposed guidelines from academic, industry, not-for-profit, and government communities.

In addition, the Food and Drug Administration (FDA) announced that it would block attempts to clone a human being because of unacceptably high risks to human subjects. Prior to pursuing a clinical investigation using cloning technology, the researcher would be required to submit an investigational new drug application (IND) to the FDA. If the agency finds that "human subjects are or would be exposed to an unreasonable and significant risk of illness or injury," that would be sufficient reason to put a study on clinical hold.[46] Currently, the FDA believes that there are unresolved safety questions pertaining to the use of cloning technology to create a human being, and until those questions are appropriately addressed in the IND, the FDA would not permit any such investigation to proceed.[47]

As stated by the U.S. Supreme Court in *Chakrabarty*, "We are told that genetic research and related technological developments may spread pollution and disease, that it may result in a loss of genetic diversity, and that its practice may tend to depreciate the value of human life."[48] However, twenty years after *Chakrabarty*, genetic research has failed to have that effect. On the contrary, genetic engineering has enhanced the value of life by providing new treatments for a vast number of diseases thereby alleviating the human pain and suffering associated with those diseases. Furthermore, development of genetically engineered crops and cattle will no doubt enhance food productivity, and new reproductive technologies will continue to permit couples to bear children who would otherwise be unable to do so.

The successful development of new therapeutics, diagnostics, reproductive technologies, and engineered crops has relied in large part on the ability to obtain patent protection for such inventions. Clearly, the issuance of patents to genetically engineered products and methods "not only results in the dissemination of technological information to the scientific community for use as a basis for further research but also stimulates investment in the research, development, and commercialization of new biologics."[49] As recognized by the Court in *Charkabarty*, "the grant or denial of patents on microorganisms is not likely to put an end to genetic research or to its attendant risks. The large amount of research that has already occurred when no researcher had sure knowledge that patent protection would be available suggests that legislative or judicial fiat as to patentablility will not deter the scientific mind from probing into the unknown."[50]

NOTES

1. See Hon. Pauline Newman, "Legal and Economic Theory of Patent Law" (Luncheon Speech to the ABA-IPL Section, 21 July 1994).

2. See Ananda M. Chakrabarty, "*Diamond* v. *Chakrabarty*: A Historical Perspective," *Principles of Patent Law*, ed. Chisum et al. (New York: Foundation Press, 1998).

3. Ian Wilmut et al., "Viable Offspring Derived From Fetal and Adult Mammalian Cells," *Nature* 382 (27 February 1997): 810–13.

4. See "Cloning Issues in Reproduction, Science, and Medicine," a report from the Human Genetics Advisory Commission and the Human Fertilization and Embryology Authority, December 1998.

5. See Eliot Marshall, "A Versatile Cell Line Raises Scientific Hopes, Legal Questions," *Science* 282 (6 November 1998): 1014–15.

6. See Sacha Waldman, "The Recycled Generation," *New York Times Magazine*, 30 January 2000.

7. Gretchen Vogel, "Capturing the Promise of Youth," *Science* 286 (17 December 1999): 2238.

8. See Cho et al., "Ethical Considerations in Synthesizing a Minimal Genome," *Science* 286 (10 December 1999): 2087–90.

9. Hutchison III et al., "Global Transposon Mutagenesis and a Minimal Mycoplasma Genome," *Science* 286 (10 December 1999): 2165–69.

10. See Frisee, *PTO Today*, available at (http://www.uspto.gov).

11. See Merz et al., "Disease Gene Patenting Is a Bad Innovation," *Molecular Diagnosis* 2 (December 1997): 299–304.

12. See *Patent, Trademark, and Copyright Journal* 59 (11 February 2000): 542–65.

13. Article 1, Section 8, of the United States Constitution.

14. 35 U.S.C. §§154, 271.

15. 35 U.S.C. §101.

16. 35 U.S.C. §§ 101, 102, and 103.

17. 35 U.S.C. §112.

18. *Diamond* v. *Chakrabarty*, 447 US 303 (1980).

19. 447 US at 308.

20. 447 US at 308.

21. *Ex Parte Hibberd*, 227 USPQ 443 (BPAI 1985).

22. *Pioneer Hi-Bred International, Inc.* v. *JEM AG Supply, Inc.*, 53 USPQ 2d 1441 (Fed Cir 2000).

23. *Ex Parte Allen*, 2 USPQ 2d (BNA) 1425, aff'd, 846 F.2d 77 (1998).

24. Commissioner of Patents and Trademarks, Policy Statement on Patentability of Animals 1077 Off. Gaz. Pat Office 24 (Apr. 7 1987).

25. To date, at least 154 patents have issued covering transgenic animals.

26. See *Moore v. Regents of the University of California*, 51 Cal 3d 120, 793 P2d 479, 15 USPQ2d 1753 (Ca 1990).

27. 15 USPQ2d at 1757.

28. *Amgen Inc v. Chugai Pharmaceutical Co*, 927 F. 2d 1200, 18 USPQ 2d (BNA) 1016 (Fed Cir 1991).

29. USC § 112, ¶ 1 dictates that an inventor describe with particularity the claimed features of an invention, to show that the inventor is in "possession" of the invention. See, e.g., *Fiers v. Sugano*, 984 F. 2d 1164, 1170-71 (Fed Cir 1993)

30. *Regents of the University of California v. Eli Lilly and Co.*, 119 F. 3d 1559, 43 USPQ 2d (BNA) 1398 (Fed Cir 1997), cert. denied, 118 S. Ct. 1548 (1998).

31. See (http://www.uspto.gov). While the Written Description Guidelines are technology neutral, the Utility Guidelines are specifically directed to biotechnology.

32. See *Patent, Trademark, and Copyright Journal* 58 (17 June 2000): 203–204.

33. See 1077 OG 24, 21 April 1987.

34. Council Directive 98/44 Art 4, 1998 O.J. (L213).

35. *Bedford v. Hunt*, 3 F Cas 37 (CCD Mass 1817).

36. *Reliance Novelty v. Dworzek*, 80 F902 (N.D. Cal. 1897).

37. *Fuller v. Berger*, 120 F 274, 275-76 (7th Cir 1903).

38. *Brenner v. Manson*, 383 US 519 (1966).

39. *E. I. du Pont Nemours Co. v. Berkely* 620 F 2d 1247, 1260.

40. *Whistler Corp v. Autotronics Inc.,* 14 USPQ 2d 1885 (ND Tex 1988).

41. *Patent Law and Policy* 36 (1992): 157.

42. Patent Nos. GB2318578 and GB2331751.

43. Patent No. EP0695351.

44. See P.L. 105-277, section 511, 112 STAT. 2681-386.

45. See (http://www.nih.gov/news/stemcell/draftguidelines.htm).

46. See Information Sheets for IRBs and Clinical Investigators, (http://www.fda.gov/oc/ohrt/irbs.html).

47. See FDA Letter, Stuart L. Nightingale, Associate Commissioner for Health Affairs (http://www.med.upenn.edu/%7Ebiothic/Cloning/fda.html).

48. *Diamond v. Chakrabarty*, 447 US 303, at 318 (1980).

49. John J. Doll, "The Patenting of DNA," *Science* 280 (1 May 1988): 689–90.

50. *Diamond v. Chakrabarty*, 447 US 303, at 318 (1980).

four

PATENTING HUMAN GENES
The Advent of Ethics in the Political Economy of Patent Law
ARI BERKOWITZ AND DANIEL J. KEVLES

J ust as the development of technology is a branch of the history of political economics, so is the evolution of patent law. The claim is well illustrated by the attempts mounted in recent years in the United States and Europe to patent DNA sequences that comprise fragments of human genes. Examination of these efforts reveals a story that is partly familiar: Individuals, companies, and governments have been fighting over the rights to develop potentially lucrative products based on human genes. The battle has turned in large part on whether the grant of such rights would serve a public, economic, and biotechnological interest. Yet the contest has raised issues that have been, for the most part, historically unfamiliar in patent policy—whether intellectual property rights should be granted for substances that are said to comprise the fundamental code of human life. The elevation of human DNA to nearly sacred status has fostered the view among many groups that private ownership and exploitation of human DNA sequences is somehow both wrong and threatening, an unwarranted and dangerous violation of a moral code.

Attempts to patent human DNA rest legally on *Diamond* v. *Chakrabarty*, the U.S. Supreme Court's decision in 1980 that allowed the patenting of living organisms modified, and hence made, by humans. The Court imposed no limits on what might be patentable, though a later ruling by the appeals board in the U.S. Patent and Trademark Office held that patents could not be obtained on a human being. A biological material in its natural state remained for the law a "product of nature" and as such was also not patentable. However, once a biological substance was isolated from the body by human artifice, it became patentable because it was no longer in its natural state. Following *Chakrabarty*, cDNA copies of complete human genes with known functions were routinely patented. Such copies do not occur naturally; they are man-made. The issue of patents on them did not break substantially new legal ground or provoke widespread reactions.

On June 20, 1991, however, J. Craig Venter and Mark D. Adams, of the National Institute of Neurological Disorders and Stroke (a division of the NIH), filed a 244-page patent application at the U.S. Patent and Trademark Office (PTO) for 315 human DNA sequences.[1] They had used commercially available automated sequencing machines to sequence random fragments of human cDNA that they took from a "library" of cDNA clones from the human hippocampus (a part of the brain). Complementary DNA (cDNA) is made from messenger RNA (mRNA); it is thus the part of DNA that figures in the creation of proteins. Such DNA is estimated to account for about 3 to 5 percent of all human DNA. It is the part of our DNA that is most likely to be useful for understanding diseases or normal functions, and is most likely to lead to lucrative products.[2] Venter, Adams, and the NIH sought patent protection for all cDNA sequences they had obtained that had no match in public DNA databases. They also published their findings in the June 21, 1991, issue of *Science* and simultaneously submitted the sequences to GenBank, a public data repository in Los Alamos, New Mexico.

What was controversial about the application was that, while Venter and Adams were applying for patent protection on only a fragment of human DNA, such patent protection seemed likely to give them control over the entire gene that the fragment identified. Venter's lab had sequenced just enough of each gene—150 to 400 base pairs, which they termed an "expressed sequence tag," or EST—to establish its unique identity.[3] Venter and Adams claimed that the ESTs would have utility "as diagnostic probes" for the presence of particular types of mRNA in specific cell types and as DNA markers for mapping locations of genes on chromosomes. They made no attempt to tie sequences to any function or disorder, or even to map the locations of most ESTs on chromosomes. Venter and his colleagues were able to churn out sequences at an unheard-of rate. Another researcher could then use the EST to locate the same cDNA from the same or another cDNA library.

Venter seemed prepared to seek patents on the vast majority of human genes, or at least a substantial fraction of the 100,000 genes estimated to be contained in the human genome. At a congressional briefing on the

Human Genome Project that summer, Venter mentioned in passing that the NIH planned to file patent applications for 1,000 such sequences a month. "I almost fell off my chair," said one briefing participant.[4] Venter and the NIH's dramatic move may have been partly a result of Venter's nononsense attitude. He later explained: "I turned twenty-one in Vietnam. So, in that situation, I saw that there is too little time in life to waste on B.S. approaches. So, in that sense, I am impatient. I want to constantly be moving forward with the discovery of new things, and I'm frustrated with how long it takes new ideas to become part of the general thinking."[5]

Venter attributed the idea for patenting the ESTs to Max Hensley, a patent attorney for the biotechnology company Genentech, who apparently suggested the idea to Reid G. Adler, the director of the NIH's Office of Technology Transfer, who in turn convinced Venter. Adler apparently felt that if the NIH could patent these DNA sequences, there would still be an incentive for companies to develop products using them because the NIH could grant the companies exclusive or partly exclusive licenses under the Federal Technology Transfer Act of 1986.[6] If the sequences were published without being patented, they would be in the public domain, and companies would thus be without that incentive to develop products from the sequence information.

Others, however, argued that the NIH patents on ESTs would inhibit industrial development of products from them.[7] Bernadine Healy, the director of the NIH, later testified: "[The] NIH is amenable to not enforcing any patent rights that may issue to partial sequences of unknown function, except in the unusual situation where the licensing of such rights is necessary to provide for the development of a therapeutic agent that might not otherwise come to market."[8] Healy also stated, "The NIH is doing this in a socially responsible way for the purposes of assuring that products that are life-saving remedies and therapies that are derived from this basic knowledge will be developed in the interest of our mission, which is science and the pursuit of health."[9] She said that it was important for the NIH to be at the table.

The debate that ensued turned in part on technical legal issues of patentability. Considerable emphasis went to the patentability of sequences that were merely fragments of genes of unknown function. Venter's approach raised questions about the "nonobviousness" and "utility" of the sequences that had not been raised by earlier gene patents. The international Human Genome Organization (HUGO) later argued: "Several uses have been suggested for genes and gene fragments to get past the utility requirement for patent protection. For any random gene, gene fragment, or collection of genes or gene fragments, it is easy to give a list of potential uses without knowledge of their true biological functions. . . . In all important cases the development of a truly useful tool for these purposes will require the investment of considerable further effort and creativity, far more than that invested in finding the initial fragment."[10]

Patent attorneys were divided over the patentability of ESTs. Steve Bent, of the firm Foley & Lardner in Washington, D.C., thought that the patent issued would be restricted to the ESTs themselves. He said, "I don't want to say that NIH is wrong, but my feeling is maybe they entered the patent process too early."[11] Max Hensley, the Genentech attorney who suggested EST patenting, nonetheless, was uncertain about its success. He said, "[T]he tummy feel to this is not quite right. You ask, 'Where's the beef?'. . . If it was 10 or 50 genes a year, I could make that fly. But when you start talking about 20,000 genes, a buzzer goes off and you wonder, How will I get that by a judge?"[12] But he also pointed out, "If these things are patentable, there's going to be an enormous cDNA arms race."[13]

Venter and the NIH responded to the legal unhappiness. On February 12, 1992, they filed a second patent application—technically a "continuation in part" (CIP) for the first application, which both added to and modified the initial application—for 2,412 ESTs in concert with publishing new EST data in the February 13 issue of *Nature*.[14] According to the CIP, their work represented "a quantum leap forward in mankind's knowledge of human gene sequences." However, they had actually eliminated several types of patent claims. This time the patent claims included the ESTs and the full genes, but not the proteins made from the genes, nor any antibodies or living cells, in contrast to the wider claims of the original patent application. (One of the NIH patent attorneys recalled, "We thought if we controlled the gene, we'd control the protein in the end anyway.") The CIP also removed the process of producing the cDNA sequences from the list of patent claims.[15] Stephen Raines, vice president for patents at Genentech, said, "It is a little dangerous to ask for the world. As the claim gets narrower, that usually helps support the argument of patentability."[16]

★ ★ ★

The utility requirement for patents has traditionally been relatively easy to meet. As Rebecca S. Eisenberg, a law professor at the University of Michigan, later wrote: "The utility requirement is rarely invoked in practice, perhaps because few people go to the trouble and expense of seeking patents on useless inventions, and no one is likely to care much if they do. . . . It is the as yet undiscovered utility of the sequences, rather than the uses that are disclosed in the patent application, that makes NIH's patent claims worth fighting about."[17]

The stakes in undiscovered utility were high, both scientifically and commercially. The NIH/Venter move divided government officials, who recognized that the issue went beyond legal technicalities to questions of political economy. While the NIH pursued the patents, James Watson, then head of the NIH's genome project, and David Galas, head of the Department of Energy's genome project, both strongly opposed the move. Watson called cDNA patenting "outrageous" and "sheer lunacy."[18] He said that "virtually any monkey" could perform this type of research. "What is important is interpreting the sequence. . . . If these random bits of sequences can be patented, I am horrified." Galas warned, "There is no coherent government policy, and we need one—quick—since the sequence is just pouring out. . . . It would be a big mistake to leave this one to the lawyers."[19] Adler claimed that the NIH would continue to pursue the patents through litigation only if industry showed interest in licenses from these patents; then, companies could pay for the legal costs related to their licenses: "It will be the company bearing the cost, not the taxpayer." He set up a November 14 meeting with industry representatives to "announce that this invention is ready for licensing."[20] (As of May 1992, one company apparently had a license application on file at the NIH, although the patent application had not yet been decided on.)[21]

Stakes in the issue were evident to interest groups outside the government. Many scientists strongly dissented from the NIH's move. Maynard Olson, a molecular geneticist then at Washington University and a member of the advisory panel to the Human Genome Project, said, "I think it's a terrible idea. . . . If the law is interpreted to give intellectual property rights for naked DNA sequences, then the law should be changed. It's like trying to patent the periodic table. . . . Patent law wasn't designed to be a kind of lottery where one guesses every large number of letters that might be the right combination."[22] George Annas, a lawyer and medical ethicist at Boston University, added, "This is not science. This is like the gold rush. That's why there are no scientists saying this is a wonderful thing."[23]

Biotech companies were also divided on the issue. The Association of Biotechnology Companies (ABC) in Washington, D.C., which represented 280 smaller companies and institutions, issued a statement in May 1992 that supported the NIH patent application, but argued that the NIH should not attempt to award exclusive licenses for the ESTs (as opposed to the full cDNAs or the proteins).[24] Lisa Raines, vice president of the Industrial Biotechnology Association (IBA), which represented 125 larger companies, including 80 percent of U.S. investment in biotechnology, noted that U.S. biotech companies support the NIH patent application "for the purpose of preserving NIH's options."[25]

However, in June 1992, an IBA committee recommended opposition to the NIH patent. The IBA committee argued that product development would require "more meaningful and costly scientific work" than the EST sequencing and that it would be "unfair to permit the Government to exercise complete control over a product to whose development the Government contributed little." In addition, the committee suggested that issuing the NIH patents would increase costs of product development as well as risks of future patent infringement litigation. Companies would be encouraged "to abandon current research efforts that are aimed at product development in favor of routine genetic sequencing for the purpose of staking claims to as much of the genome as possible."[26] Richard Godown, president of the IBA, observed, "If somebody spends a lot of time and money to discover the whole gene and its function, and then discovers they've got to deal with somebody who owns a patent to part of it, suddenly the commercial possibilities become clouded."[27] Moreover, the Pharmaceutical Manufacturers Association, which had one hundred member companies, actively opposed the NIH patent in a May 28, 1992, letter to Louis W. Sullivan, the secretary of health and human services, saying that "a governmental policy of ownership and licensing of gene sequences would inevitably impede the research and development of new medicines in this country."[28]

In August 1992, Adler conceded that "it is clear that the trade associations are not interested in exclusively licensing sequences of unknown function, assuming that they are patentable, as an incentive for product development."[29] In fact, Healy had met privately with a group of industry

CEOs to get their input even before she gave the go-ahead for the original Venter application. At that meeting, the CEO of a large company told Healy that the industry was not concerned about individual scientists filing for EST patents, but it did not want the government to have them. This CEO apparently felt that patent rights could be easily acquired from a scientist, in exchange for an appropriate sum of money, but that getting them from the U.S. government would be more problematic. The CEO's views made Healy all the more determined to stick with the NIH position.[30]

Foreign governments were apprehensive that their biotechnology enterprises would be competitively disadvantaged by the NIH patents on ESTs. Government officials in France, Italy, and Japan announced their countries' opposition to such patents early on.[31] The French Academy of Sciences issued a statement on January 13, 1992, condemning "any measure which, answering purely to a logic of industrial competition, strove to obtain the legal property of genetic information data, without even having taken care to characterize the genes considered."[32] In March 1992, the British retaliated against the NIH: the minister of science, Alan Howarth, announced that the Medical Research Council (MRC) would also seek cDNA patents. He said, "a decision by the U.K. Medical Research Council (MRC) not to seek patents when researchers funded by public bodies in other countries have or may do so could place the U.K. at a relative disadvantage."[33] However, Howarth also claimed that the United Kingdom was attempting to negotiate an international agreement not to patent "genome sequences of unknown utility identified as a result of publicly funded research."[34]

In July 1992, the MRC applied to the European Patent Office (EPO) and the U.S. Patent and Trademark Office (PTO) for patents on about 1,200 ESTs, all the while claiming that it opposed EST patenting.[35] (In October 1993, the MRC announced that it would not apply for any additional EST patents. David Owen, the MRC's director for technology transfer, explained: "By filing a patent, we felt that we would get a seat at any table where the issue was discussed. We've now got places in the most important discussions on the patent issue—in the patent offices of the world and in a public inquiry for the U.S. Congress by its Office of Technology Assessment.")[36] In April 1992, James Watson resigned as head of the NIH genome project, pointing to the NIH's attempts to patent ESTs.[37] (Watson, however, was also under fire for conflict of interest: he owned shares in a biotech company and refused to sell them. The attack on Watson for conflict of interest was led by Frederick Bourke, a businessman who aimed to form a private company to sequence the nematode genome—a project that the NIH was already pursuing in collaboration with the British.)[38]

In the meantime, Venter himself, along with three dozen other scientists at a human genome meeting in Brazil in May 1992, signed a resolution that opposed the patenting of "naturally occurring gene sequences," while supporting the patenting of specific uses for sequences. Venter said he supported the NIH patent application only because it stimulated debate; he hoped the patent would not be issued.[39] (Venter, however, did stand to gain from the patent. Under the terms of the Federal Technology Transfer Act of 1986, he and Adams would be personally entitled to at least 15 percent of any royalties accrued from licensing of patents, with the majority of such royalty income designated for funding of research in Venter's laboratory.)[40]

In July 1992, Venter announced that he was leaving the NIH to head a new private, nonprofit research center called The Institute for Genomic Research (TIGR). TIGR received $70 million as a ten-year grant from a New Jersey venture capital group called Healthcare Investment Corporation, which had also created several biotech companies, including Genetic Therapy Inc. The chair of Healthcare Investment Corporation, Wallace Steinberg, asserted that American scientists needed to patent genes before their European and Japanese competitors beat them to it: "I suddenly said to myself that if this thing doesn't get done in a substantive way in the United States, that is the end of biotechnology in the United States."[41] While TIGR itself would be nonprofit, a new biotech company, called Human Genome Sciences Inc. (HGS), was created to develop and market products resulting from TIGR's research. TIGR and the new company would be set up in Montgomery County, Maryland, near the NIH.[42] Venter took thirty NIH researchers with him and said TIGR would "do the genome project," beginning with a scaled-up continuation of his project to sequence random ESTs.[43] He predicted that TIGR would discover "a majority of human genes within the next 3 to 5 years at a pace of up to 1000 genes per day."[44] Even critics wanted to see the NIH EST patent application legally decided. "We need a definitive answer," Paul Berg said. "Withdrawing the patent would resolve nothing."[45]

On August 20, 1992, the expedited decision of the U.S. PTO was announced. The initial review concluded that the patent claims were "vague, indefinite, misdescriptive, inaccurate, and incomprehensible."[46] The

PTO rejected the patent claims on the basis of each of their major criteria; the claims did not meet their standards for utility, "enablement," "nonobviousness," or novelty. (In contrast, most of the public criticism of the patent application had focused exclusively on utility as the problematic criterion.) The PTO claimed that the sequences were obvious for a surprising reason. Molecular geneticists had traditionally regarded a fifteen-base pair sequence (or "15mer") as the minimum length of DNA sufficient to identify a gene. (However, many researchers already had come to believe that a longer stretch of DNA was, in fact, necessary for this purpose.)[47] The PTO examiners searched through conventional databases of DNA sequences and found that randomly selected 15mers from the Venter application sometimes occurred within published sequences of human or non-human DNA. Hence, they concluded, "It would be obvious for someone ordinarily skilled in the art" to find Venter's ESTs using one of these published 15mers as a probe.[48] Thus, ironically, while most of the scientific and biotech community was urging that the patents be denied because short fragments of genes with completely unknown functions should not be patentable, and if patented, might hinder the elucidation of full gene sequences and functions, the PTO effectively argued that previous publication of even smaller fragments had already undermined the patentability of Venter's fragments.

The PTO also concluded that the sequences were not novel because the DNA was taken from a commercially available "library." Bernadine Healy, the director of the NIH, objected during the gene patent hearings organized in September by Sen. Dennis DeConcini (D–Ariz.), chair of the Senate Subcommittee on Patents, Copyrights, and Trademarks: "Taken to its logical extension . . . the PTO's reasoning would deny novelty to virtually all products isolated from expected sources of biomolecules in nature, such as blood, saliva, or tissues."[49] Craig Venter, also testifying at the hearings, argued that patent law should be changed to permit future patenting of DNA sequences that contain previously published fragments.[50] Healy agreed: "The PTO's position here suggests the need to at least consider a legislative remedy and international agreement that prior publication of partial gene sequences not preclude a subsequent patent on the full genes and/or partial genes with known function."[51] Sen. Pete Domenici (R–N.M.) planned to introduce such a bill.[52]

★ ★ ★

The door to ethical debates over gene patenting was opened in the Senate by
Mark O. Hatfield (R–Oreg.), who had already introduced a bill to impose a
five-year moratorium on patenting animals. Although the bill did not refer to
gene patenting, Hatfield argued in speeches that gene patenting raises the
"specter of removing the building blocks of life from the common possession
of us all and shifting them to the private use and profit of researchers or corporations." He also complained that biotechnology generally has reduced man
to a "biological machine."[53] Hatfield's bill, reconstituted as an amendment to
the NIH reauthorization bill, providing a three-year moratorium on patenting
of both living organisms and "genetic matter," did not advance. Hatfield withdrew the amendment after reaching an agreement with Sens. Dennis
DeConcini and Edward M. Kennedy (D-Mass.), chairs of the Senate Committee on Labor and Human Resources, that they would each schedule hearings on gene patenting. Hatfield, DeConcini, and Kennedy also requested a
report from the Office of Technology Assessment (OTA) on legal, ethical, and
economic issues raised by human gene patenting.[54]

Ethical considerations had been advanced earlier—by critics of genetic
engineering such as Jeremy Rifkin and his allies as well as by a variety of
clerics, a number of whom had been recruited by Rifkin—in connection
with the *Chakrabarty* decision and then with the patenting of animals. They
contended that the patenting of life "desacralized" it, reducing it to a mere
commodity, like tennis balls. Now, in hearings in September 1992, the issue
of patenting DNA sequence fragments prompted a reconsideration in an
ethical framework of whether human genes should be patented at all.

Andrew Kimbrell, the policy director and attorney for Jeremy Rifkin's
Foundation on Economic Trends, argued in favor of a moratorium on gene
patenting such as Senator Hatfield had suggested, saying, "We are right in
the middle of an ethical struggle on the ownership of the gene pool." Kimbrell reviewed innovations in the patenting of biological entities since the
1980 *Chakrabarty* Supreme Court decision, which included the patenting of
human cells and of entire mammals. Referring to these innovations as the
"the children of *Chakrabarty*," he suggested that the *Chakrabarty* decision be
effectively reversed by legislation: "We need Congress to intercede to decide
where this ethical and legal free-fall ends." Before *Chakrabarty*, "[w]e never
allowed the patenting of animals. It was tried. They tried to patent hybrid

chickens and the answer was no. We have never allowed nature itself to be patented because that is our common heritage."[55] Kimbrell added:

> There is little question that, unless stopped, the patenting juggernaut will continue to transgress into life in all its forms. As research continues in cell analysis and in the deciphering of the human genome, corporations and researchers will fight for patent ownership of commercially valuable genes and cells held to be the key to health, intelligence or youth. As there are advances in reproductive technologies human embryos may be up for patent grabs. Animals with increasing number of human genes will be patented. Genetically engineered human body parts will almost certainly be patented. And looking into the more distant future perhaps a genetically altered human body itself may be patentable. As patenting continues, the legal distinction between life and machine, life and commodity will begin to vanish.[56]

However, Kimbrell was in a minority at the hearings. He was strongly opposed by patent attorneys and by representatives of the biotech industry. For example, Genentech vice president David Beier, testifying as a representative of the Industrial Biotechnology Association (IBA) and the Association of Biotechnology Companies (ABC), opposed a moratorium on human DNA patenting.[57] A key criticism came from William D. Noonan, a physician and patent attorney who testified on behalf of the Oregon Biotechnology Association. Noonan insisted on distinguishing between issues of political economics and issues of ethics. The former had a place in disputes over patent policy; the latter, at least in the United States, did not, even though they might be legitimate in principle. Noonan argued that because of advances in human genetics, "we have to confront some of the darker questions about our human nature as we gain the power to practice eugenics on a scale and with a precision that was previously impossible." In this context, however, the debate about patenting human ESTs was a red herring. Noonan elaborated:

> [T]here is nothing inherently wrong or even ethically new about patenting DNA molecules. We have been patenting chemical components of the human body for years. Patents have been issued for decades on purified proteins, enzymes, neuropeptides, and many other gene products. There is no inherent ethical distinction between patenting these

molecules and a purified molecule of DNA. Promoting the development
of new medical treatments ethically justifies gene patents. Patent applications have been filed in recent years on genes involved in cystic fibrosis,
neurofibromatosis, Fanconi's anemia, and other diseases. These filings did
not provoke the ethical outcry. It was only when the NIH filed Dr.
Venter's patent applications on cDNAs of unknown function that a sus-
tained international debate arose about the ethics of gene patents. I think
we run the risk of failing to address the real ethical concerns if we
become too fixated on what is essentially a problem in international scientific politics and the uncertainty about the scope of patent law. What
we should instead talk about is the social impact of human genome
research. Do we want to practice molecular eugenics on humans and animals, and what is the acceptable scope of such eugenic efforts? These ethical questions have nothing to do with patent law and cannot be
addressed by changing the scope of patentable subject matter. The Patent
Office is, of course, the wrong place to conduct any ethical inquiry.[58]

In support of biotechnology and in opposition to a moratorium, Sen.
Orrin Hatch (R-Utah) warned against any measure that jeopardized jobs in
biotechnology. Such a measure would "certainly undermine our world com-
petitiveness," he warned, asking, "Do any of my colleagues believe the Euro-
peans and Japanese are going to slow down their efforts, let alone engage in
a moratorium in this cutting edge industry? Of course not. They are going
to take advantage of it."[59] Domenici chimed in that gene patenting would
assist the search for cures and treatments for genetic disease.[60]

Bernadine Healy also opposed a moratorium on gene patenting, saying
it would be contrary to the Federal Technology Transfer Act of 1986 and
would be the "death knell for the patent system in the biotech field."[61] Healy
argued that the attempt to patent human gene fragments did not raise ethical
concerns as serious as the patenting of entire organisms, which she personally regarded as questionable, but which had already received the stamp of
approval of the U.S. Supreme Court. She also pointed out that complete
human gene sequences that code for proteins with known functions had been
routinely patented without a fuss, and added: "Some have said that we should
not patent our universal heritage as a matter of ethics. But the same people
who are saying that—the French Government has said this many times— . . .
[also say that] if you know the function of the gene and it has commercial
value, then you can patent it. That seems to be an ethical double standard."[62]

★ ★ ★

Healy testified that the NIH ought to pursue the appeals process, because "truncating the patent process before it yields useful information would leave uncertainty and would perpetuate a policy void in patent law in the technology transfer system."[63] The NIH did seek approval for an appeal. Healy later asserted that the NIH's "outside patent attorney" was "very optimistic that most of the concerns raised in the preliminary finding can be met." She added: "I don't think it's a question of winning or losing. It is a question of resolving uncertainty that is unsettling to both the scientific community and to the policy makers. . . . It will ultimately be a decision made by the Secretary [of health and human services (HHS)] as to whether or not we will continue these patent proceedings."[64] In late 1992, during the presidential transition from the administration of George H. W. Bush to that of William Clinton, HHS approved the NIH appeal, which was submitted to the PTO on February 19, 1993. This amended application, however, was also rejected by the PTO, on August 10, 1993.[65]

The NIH might then have appealed the application in the courts. However, Healy was soon succeeded in the directorship of the NIH by Harold Varmus. In February 1994, Varmus announced that the NIH was withdrawing its patent application on all ESTs (the number of ESTs in the application had since increased to 6,869), saying that such patents are "not in the best interests of the public or science."[66] Varmus's decision was heavily influenced by advice from Rebecca Eisenberg, who served on an advisory panel.[67] David Galas, having moved from the Department of Energy to Seattle to become scientific director of a company called Darwin Molecular, thought that the NIH was right to drop the application: "If [NIH officials] didn't think that the granting of the patents was in the public interest, then they were put in the position of pursuing with public funds something they hoped they'd lose."[68] The British Medical Research Council (MRC) soon followed suit, withdrawing its own EST patent applications.[69]

The Venter/NIH application, however, had let the genie out of the bottle, perhaps irreversibly. Several companies had followed the NIH's lead and filed their own patent applications on ESTs. Venter initially claimed that neither his nonprofit research institute (TIGR) nor its associated company (HGS) would file patent applications for ESTs with unknown function.[70] A subsequent company prospectus for HGS, however, indicated that the company had filed patent applications for 9,900 ESTs.[71] In addition, Incyte Pharmaceuticals Inc., of Palo Alto, California, filed for more than 40,000 ESTs.[72] Incyte planned to file applications for as many as 100,000 ESTs each year.[73] These applications were not affected by the change of heart at the NIH.

The genie was out on gene patenting as such, too. In May 1995, prompted by Jeremy Rifkin's Foundation on Economic Trends, several clerics held a press conference in Washington, D.C., to announce that a coalition of 180 religious leaders representing eighty denominations and Rifkin's group had signed a Joint Appeal against Human and Animal Patenting that also opposed the patenting of any human genes. The group included the executive director of the American Muslim Council, members of the United Methodist Church, the secretary general of the Reformed Church in America, a member of the Christian Life Commission of the Southern Baptist Convention, and the director of the Religious Action Center of Reform Judaism. Rifkin said, reiterating one of his longstanding themes, "By turning life into patented inventions, the government drains life of its intrinsic nature and sacred value." And one of the Baptist clerics told the *New York Times* that "altering life forms, creating new life forms [is] a revolt against the sovereignty of God and an attempt to be God."[74] He and a fellow Baptist later added that "the right to own one part of a human being is *ceteris paribus* the right to own all the parts of a human being. This right must not be transferred from the Creator to the creature."[75]

★ ★ ★

All the while, the biotechnology industry on both sides of the Atlantic had been paying close attention to the patentability of genes in Europe, and there, by law, ethical issues enjoyed a seat at the table of patent policymaking. According to article 53a of the European patent convention, patents were not allowable that violated "public order and morality." Since the late 1980s, the European Commission, the executive arm of the evolving European Community, had been seeking to promulgate a directive establishing the patentability of biotechnological inventions. It had been repeatedly blocked by the European Parliament, where ethical opponents—Green party members, for example—were strong. A new compromise draft directive from the commission was before Parliament in 1994 that allowed the patenting of human genes provided that "they cannot be linked to a specific individual."[76] Willy Rothley, of Germany, the head of the effort to find suitable language in Parliament, considered the language a success: "The European Parliament has been able to impose an ethical dimension on patent rights and has been able to obtain most of the guarantees that it was asking for."[77]

However, Linda Bullard of the European Green Party countered: "We feel that Parliament, having voted previously against patents on parts of the human body—including genes—under any circumstances, is morally obliged to reject this compromise. This is not a question of individual human dignity, but of collective human dignity."[78] Like William Noonan, in his congressional testimony in 1992, researchers and biotech leaders, especially in the United States, argued that morality had no place in the discussion about patenting DNA. George Poste, research director of SmithKline Beecham, argued typically: "Patent law is entirely unsuited to arbitrate on moral and ethical questions. A ban on genomic patents will not shield society from the evolution of genetic medicine or the need for vigilance against the misuse of genetic testing or the genetic modification of humans."[79]

The compromise directive, however, died in March 1995, when the European Parliament voted 240 to 188 against approval, with 23 abstentions. This signaled the end of the directive, at least temporarily, because under the terms of the Maastricht Treaty, the European Parliament has an effective veto over the process. Rifkin declared, "It is a great victory."[80]

In the United States in May 1996, Rifkin spoke out on behalf of a group of women's rights leaders against patenting human genes implicated in breast cancer. He claimed that efforts to patent these genes represented an "assault on women" and "denies them control over the most intimate aspect of their being, their bodies' genetic blueprint."[81] He announced that a new coalition would file a petition with the PTO to challenge the patent claims filed by Myriad Genetics Inc. of Salt Lake City, Utah, on BRCA1 and BRCA2, the genes that make women susceptible to breast cancer, and that its associate, Mark Skolnick of the University of Utah, had isolated. Rifkin's statements were endorsed by members of women's health organizations in sixty-nine countries, including author Betty Friedan, Gloria Steinem (the consulting editor of *Ms.* magazine), and Bella Abzug, the cochair of the Women's Environment and Development Organization.[82] Abzug, who was also a former U.S. congresswoman and a breast cancer survivor, said, "Human genes are not for sale or profit. Any attempt to patent human genetic materials by individuals, scientific corporations, or other entities is unacceptable."[83] Carolyn A. Marks, a breast cancer and ovarian cancer patient and a member of the National Ovarian Cancer Coalition, said it "boggles [her] mind" that someone would claim a patent on a human gene.[84] She added that the claim gave "a new definition to 'chutzpah,'" because Myriad was seeking a patent "for something that, in essence, was already there."[85] Rifkin called the new campaign "the beginning of the genetic rights movement around the world."[86] Carl Feldbaum, president of the Biotechnology Industry Organization, noted that the previous year Rifkin had "wrapped the gene patenting issue in clerical garb. This year he came out [in] feminist garb."[87]

The more numerous the number of genes for human disease that became known, the larger the number of interest groups whom Rifkin might enlist in the anti-gene-patenting cause. But the American biotechnology industry constituted a formidable pressure group in both the United States and Europe against permitting the kind of ethical issues raised by Rifkin, the European Greens, and their allies to figure consequentially in the formation of patent policy for living organisms. The stakes for the biotechnology industry were high. It would obviously be costly to make what is patentable in the United States unpatentable in Europe. American biotechnologists also had a growing number of European allies. Interpharma, an association representing Swiss pharmaceutical companies, supported patents on genes or gene fragments "in a form that does not occur in nature."[88] In Germany biotechnology had been lagging. By 1996, the political winds had shifted and the federal government began offering financial incentives to encourage German biotechnology.[89] As Maria Leptin, head of the genetics faculty at the University of Cologne, put it: "[I]f there's anything that's more important [to Germans] than saving the environment, it's saving jobs. As soon as people saw the [pharmaceutical] industry possibly disappearing, morality went out the window."[90]

In summer 1997, the European Parliament took up for reconsideration the question of the patenting of biological inventions, a policy issue that it had been dealing with for about a decade. In spring 1998, it approved, as part of a broad directive on the subject, the patenting of genes that had been fully characterized. But, for the first time anywhere, the policy on biotechnology for the European Union imposed specific ethical constraints on what can be patented. It opted for the encouragement of biotechnology while adopting ethical restrictions to address the apprehensions of the opposition. Holding that biotechnology patents must safeguard the dignity and integrity of the person, the directive prohibited patents on, among other entities, human parts, human embryos, and the products of human cloning. It also forbade the grant of patents on animals if what they suffer by being modified (say, genetically) exceeds the benefits (say, medical) that the modification would yield.[91]

In the United States in December 1997, Jeremy Rifkin and a biologist announced that, as a provocation, they would seek a patent on methods to create a human/animal hybrid, a creature part animal and part person.[92] Bruce Lehman, the U.S. commissioner of patents, declared that the Patent and Trademark Office would, in general, reject patents that were "injurious to the well-being, good policy, or good morals of society." Patent lawyers roundly attacked Lehman, contending that he had no authority in U.S. patent law, because it is literally amoral to back such a prohibition.[93] Yet even if ethics has no rightful presence in American patent policy, an ethical principle—that the human genome must not be locked up—has been creeping into it through the issue of gene patenting. And nothing has done more to introduce it than the robust ambitions of Craig Venter.

In May 1998, Venter announced that he would leave the nonprofit TIGR to move to a new, for-profit company, called Celera, that would be located next door, in Rockville, Maryland. Celera would aim to sequence all the DNA in the human genome by 2001, using rapid new automated machines supplied by its principal owner, the Perkin-Elmer Corporation.[94] Venter declared that Celera would make all its sequence data publicly available while at the same time earn money from selling access to the information.[95] Venter's rapid-fire approach to the sequencing prompted scientific critics to predict that his company's data would contain numerous serious gaps in the DNA.[96] It was also unclear how the company could publish and profit from its sequence data. Early in 2000, strategies that Celera said it would follow to profit from its work appeared to threaten broad access to the sequence information.[97] Indeed, suspicion of Celera's intentions appeared to prompt President Clinton and British Prime Minister Tony Blair to declare in early 2000 that information about human DNA sequences should be released into the public domain.[98]

But Venter has revived his original goal of wholesale gene patenting.

Along with several other genomic companies, Celera has proposed to use ESTs to identify new genes and guess their function by finding genes of known function and similar structure through computerized searches of the genomic data base. The company would then seek utility patents covering these new genes, arguing that their functions were likely the same as those of the genes with similar structures.[99] That strategy stimulated a forceful statement in late March 2000 by Aaron Klug and Bruce Alberts, the presidents, respectively, of the Royal Society of London and the National Academy of Sciences in the United States. They called guessing at gene function by computerized searches of genomic data bases "a trivial matter." Its outcome might satisfy "current shareholders' interests," but it did "not serve society well." Holding that its results did not warrant patent protection, they stressed that "the human genome itself must be freely available to all humankind."[100]

The U.S. Patent and Trademark Office, however, flatly disagrees, judged by its current policy on the patenting of genes and DNA sequences. At the end of 1999, it invited public comments on that policy and subsequently received them from thirty-five individuals and seventeen organizations. Some of the comments were ethical, echoing those of Alberts and Krug; some were legal or practical, raising objections, for example, to granting patents on DNA sequences such as ESTs by arguing that they should not be patentable because they exist in nature. In January 2001 the office found reasons to refuse to incorporate any of the comments in its policies. Indeed, its responses to the comments in effect promulgated a policy governing the patentabilty of genes and DNA sequences that is enormously broad.[101]

Gene patenting has exposed a conflict and, possibly, an incompatability in patent policy between the United States and the European community. Even though the former does not impose ethical constraints on the patentability of products, the latter does, with the consequence that what may be patentable in the United States may not be so in Europe. Paradoxically, while trade barriers have been steadily falling with globalization, at least in the commerce of living organisms and their parts, patent barriers may be arising to some degree. The transatlantic mismatch aside, within both the United States and Europe, gene patenting has prompted important challenges to the scope of intellectual property rights in genes. Given that the human genome is widely regarded as a common birthright of people everywhere, governments may feel increasing pressure to limit the property rights sought in DNA sequences.

NOTES

1. Christopher Anderson, "US Patent Application Stirs up Gene Hunters," *Nature*, 353 (10 October 1991), 485–86; J. Craig Venter and Mark Adams, "Sequences," U.S. Patent and Trademark Office Serial Number 07/716,831, 20 June 1991.

2. Mark D. Adams, Jenny M. Kelley, Jeannine D. Gocayne, Mark Dubnick, Mihael H. Polymeropoulos, Hong Xiao, Carl R. Merril, Andrew Wu, Bjorn Olde, Ruben F. Moreno, Anthony R. Kerlavage, W. Richard McCombie, and J. Craig Venter, "Complementary DNA Sequencing: Expressed Sequence Tags and Human Genome Project," *Science* 252 (21 June 1991): 1651–56.

3. Ibid.

4. Leslie Roberts, "Genome Patent Fight Erupts," *Science*, 254 (11 October 1991), 184–86.

5. Karen Young Kreeger, "Genome Investigator Craig Venter Reflects on Turbulent Past and Future Ambitions," *Scientist* 9 (24 July 1995), 1.

6. Federal Technology Transfer Act of 1986, Public Law 99-502 (20 October 1986).

7. Roberts, "Genome Patent Fight Erupts."

8. Hearing before the Senate Subcommittee on Patents, Copyrights and Trademarks of the Committee on the Judiciary, *The Genome Project: The Ethical Issues of Gene Patenting*, 2:S. Hrg. 102–1134, 22 September 1992 (U.S. Govt. Printing Office, 1993).

9. Ibid.

10. The Human Genome Organization, "HUGO Statement on the Patenting of DNA Sequences," January 1995.

11. Roberts, "Genome Patent Fight Erupts."

12. Ibid.

13. Anderson, "US Patent Application Stirs up Gene Hunters."

14. J. Craig Venter et al., "Sequences Characteristic of Human Gene Transcription Product," U.S. Patent and Trademark Office serial number 07/837,195, *Biotechnology Law Report* 11 (May–June 1992): 260–317; M. D. Adams et al., "Sequence Identification of 2,375 Human Brain Genes," *Nature* 357 (4 June 1992): 367–68; Andy Coghlan, "US Gene Plan 'Makes a Mockery of Patents,'" *New Scientist* 133 (22 February 1992): 10.

15. Gillis, "The Patent Question of the Year"; interview (by Berkowitz) with Robert Benson, 2 April 1997.

16. Leslie Roberts, "NIH Gene Patents, Round Two," *Science* 255 (21 February 1992): 912–13.

17. Rebecca S. Eisenberg, "Genes, Patents, and Product Development," *Science* 257 (14 August 1992): 903–908.

18. Anderson, "US Patent Application Stirs up Gene Hunters"; Roberts, "Genome Patent Fight Erupts."

19. Roberts, "Genome Patent Fight Erupts."

20. Ibid.

21. Gillis, "The Patent Question of the Year."

22. Anderson, "US Patent Application Stirs up Gene Hunters."

23. D'Arcy Jenish, "A Patent on Life: Scientists Seek Legal Rights to Genes," *Maclean's* 105 (31 August 1992): 38–39.

24. Association of Biotechnology Companies, "ABC Statement on NIH Patent Filing for the Human Genome Project," *Biotechnology Law Report* 11(September–October 1992): 578–96.

25. Reid G. Adler, "Genome Research: Fulfilling the Public's Expectations for Knowledge and Commercialization," *Science* 257 (14 August 1992): 908–14; Eisenberg, "Genes, Patents, and Product Development"; Roger S. Johnson, "NIH Files for Second Patent on Expressed Sequence Tags," *Genetic Engineering News*, 1 March 1992, 17.

26. Eisenberg, "Genes, Patents, and Product Development."

27. D'Arcy Jenish, "A Patent on Life."

28. Adler, "Genome Research"; Eisenberg, "Genes, Patents, and Product Development."

29. Adler, "Genome Research."

30. Bernardine Healy, telephone interview with Ari Berkowitz.

31. Norton D. Zinder, "Patenting cDNA 1993: Efforts and Happenings," *Gene* 135 (15 December 1993), 295–98.

32. Academy of Sciences, Paris, *The Patentability of the Genome*, Academy of Sciences, Paris Bilingual Report No. 32 (Lavoisier: Paris, 1995).

33. Gillis, "The Patent Question of the Year."

34. Alan Howarth, "Patenting Complementary DNA," *Science* 256 (3 April 1992): 11.

35. Andy Coghlan, "Truce Declared in Gene Patent War," *New Scientist* 140 (6 November 1993): 10; Christopher Anderson, "NIH Drops Bid for Gene Patents," *Science* 263 (18 February 1994): 909–10.

36. Coghlan, "Truce Declared in Gene Patent War."

37. Michael Waldholz and Hilary Stout, "A New Debate Rages over the Patenting of Gene Discoveries," *Wall Street Journal*, 17 April 1992, 1.

38. Jerry E. Bishop, "At Center of Human Genome Project, Nobel Laureate and Businessman Clash," *Wall Street Journal*, 13 April 1992, B6.

39. Anderson, "US to Seek Gene Patents in Europe."

40. Federal Technology Transfer Act of 1986, Public Law 99-502 (20 October 1986).

41. Jenish, "A Patent on Life."

42. Christopher Anderson, "Controversial NIH Genome Researcher Leaves for New $70-million Institute," *Nature* 358 (9 July 1992): 95.

43. Jenish, "A Patent on Life"; Anderson, "Controversial NIH Genome Researcher Leaves."

44. Senate Subcommittee, *The Genome Project*.

45. Roberts, "NIH Gene Patents, Round Two."

46. James Martinell, U.S. Patent and Trademark Office, Art Unit 1805, Examiner's Action on Venter et al. patent application, no. 07/837,195, 20 August 1992, *Biotechnology Law Report* 11(September–October 1992): 578–96.

47. Christopher Anderson, "NIH cDNA Patent Rejected: Backers Want to Amend Law," *Nature* 359 (24 September 1992): 263.

48. Martinell, "Examiner's Action."

49. Senate Subcommittee, *The Genome Project*.

50. Anderson, "NIH cDNA Patent Rejected: Backers Want to Amend Law."

51. Gillis, "Ethics Muddle NIH's Genome Patent Application," *BioScience* 42 (December 1992): 874.

52. Anderson, "NIH cDNA Patent Rejected," 263.

53. Roger S. Johnson, "Gene Patents," *Genetic Engineering News*, 15 June 1992, 3, 20.

54. Gillis, "Ethics Muddle NIH's Genome Patent Application"; Senate Subcommittee, *The Genome Project*.

55. Senate Subcommittee, *The Genome Project*.

56. Ibid.

57. Ibid.

58. Ibid.

59. Ibid.

60. Ibid.

61. Ibid.

62. Ibid.

63. Ibid.

64. Ibid.

65. J. Craig Venter et al., Amendment to "Sequences Characteristic of Human Gene Transcription Product," U.S. Patent and Trademark Office no. 07/837,195, 19 February 1993; James Martinell, U.S. Patent and Tradmark Office, Art Unit 1805, Examiner's Action on Venter et al. patent application no. 07/837,195, 10 August 1993.

66. Christopher Anderson, "NIH Drops Bid for Gene Patents," *Science* 263 (18 February 1994): 909–10.

67. Ibid.

68. Ibid.

69. Ibid.

70. Senate Subcommittee, *The Genome Project.*

71. Anderson, "NIH Drops Bid for Gene Patents."

72. Ibid.

73. Anderson, "NIH to Appeal Patent Decision."

74. Ted Peters, "Patenting Life: Yes," *First Things: A Monthly Journal of Religion and Public Life* 63 (May 1996): 18–20.

75. Richard D. Land and C. Ben Mitchell, "Patenting Life: No," *First Things* 63 (May 1996): 20–22.

76. David Dickson, "British MPs 'Likely to Oppose Gene Patents,'" *Nature* 373 (16 February 1995): 550.

77. Ibid.

78. Ibid.

79. George Poste, "The Case for Genomic Patenting," *Nature* 378 (7 December 1995): 534–36.

80. David Dickson, "European Parliament Rejects Bid to Stem Confusion over Gene Patents," *Nature*, 374 (9 March 1995): 103.

81. "US Coalition Counters Breast Gene Patents," *Nature* 381 (23 May 1996): 265.

82. Eliot Marshall, "Rifkin's Latest Target: Genetic Testing," *Science* 272 (24 May 1996), 1094; "US Coalition Counters Breast Gene Patents," 265.

83. Marshall, "Rifkin's Latest Target: Genetic Testing."

84. Ibid.

85. Adam Marcus, "Owning a Gene: Patent Pending," *Nature Medicine* 2 (July 1996): 728–29.

86. Marshall, "Rifkin's Latest Target."

87. Marcus, "Owning a Gene."

88. David Dickson, "European Patent Directive in Critical Test over Genes," *Nature* 372 (24 November 1994): 310.

89. Steven Dickman, "Germany Joins the Biotech Race," *Science* 274 (29 November 1996): 1454–55.

90. Ibid.

91. European Community, "Directive 98/44/EC of the European Parliament and of the Council of 6 July 1998, On the Legal Protection of Biotechnological Inventions," *Official Journal of the European Communities* (30 July 1998): L213/13–21.

92. David Dickson, "Legal Fight Looms over Patent Bid on Human/Animal Chimaeras," *Nature* 392 (2 April 1998): 423.

93. Ibid.; Meredith Wadman, ". . . As US Office Claims Right to Rule on Morality," *Nature* 393 (21 May 1998): 200.

94. Eliot Marshall and Elizabeth Pennisi, "Hubris and the Human Genome," *Science* 280 (15 May 1998), 994–95; J. Craig Venter et al., "Shotgun Sequencing of the Human Genome," *Science* 280 (5 June 1998): 1540–42.

95. J. Craig Venter et al., "Shotgun Sequencing of the Human Genome."

96. Marshall and Pennisi, "Hubris and the Human Genome."

97 Eliot Marshall, "Talks of Public-Private Deal End in Acrimony," *Science* 287 (10 March 2000): 1723–25.

98. Declan Butler, "US/UK Statement on Genome Data Prompts Debate on 'Free Access,'" *Nature* 404 (23 March 2000): 324–25.

99. Professor Rebecca Eisenberg, University of Michigan Law School, conversation with author, 20 March 2000.

100. Bruce Alberts and Sir Aaron Klug, "The Human Genome Itself Must be Freely Available to All Humankind," *Nature* 404 (23 March 2000): 325.

101. Department of Commerce, Patent and Trademark Office, "Revised Utility Examination Guidelines: Request for Comments," *Federal Register* 64, no. 244, 21 December 1999, 71440; Department of Commerce, Patent and Trademark Office, "Utility Examination Guidelines," *Federal Register* 66, no. 4 (5 January 2001): 1092. I am indebted to Professor Hal Edgar, Columbia University Law School, for calling my attention to these documents.

five

DISCOVERIES
Are There Limits on What May Be Patented?
JON F. MERZ

A re there limits on what may be patented? The intent of Congress in enacting the patent law is to protect for a limited time "anything under the sun that is made by man."[1] But the courts have tried to fashion a limitation on what may be patented, holding that certain things discovered in the world around us may not be removed from the common resource available to all. As the Supreme Court described the "product of nature" doctrine, "[t]he laws of nature, physical phenomena, and abstract ideas have been held not patentable. . . . Thus a new mineral discovered in the earth or a new plant found in the wild is not patentable subject matter. Likewise, Einstein could not patent his celebrated law that $E = mc^2$; nor could Newton have patented the law of gravity."[2]

One judge reviewed the cases on the products of nature and concluded, "The common thread throughout these cases is that claims which directly or indirectly preempt natural laws or phenomena are proscribed, whereas claims which merely utilize natural phenomena via explicitly recited manufactures, compositions of matter, or processes to accomplish new and useful end results define statutory inventions."[3] Such statements do little to help resolve specific cases where the line blurs between what is natural and occurs in nature and products or processes used by humankind to useful ends. One such case involves disease genes.

In a recent article, several colleagues and I criticized disease gene patenting. Disease gene patents claim exclusivity to the diagnosis of disease by identifying genetic mutations which have been associated with the occurrence of or risk of the disease.[4] We argued that patents such as this, which are rapidly increasing in number as new gene–disease relationships are found, violate the product of nature doctrine.

An example is the patent for a gene that is predictive of breast and ovarian cancer risks, called BRCA1. The patent claims:

> 1. A method for screening germline of a human subject for an alteration of a BRCA1 gene which comprises comparing germline sequence of a BRCA1 gene or BRCA1 RNA from a tissue sample from said subject or a sequence of BRCA1 cDNA made from mRNA from said sample with germline sequences of wild-type BRCA1 gene, wild-type BRCA1 RNA, or wild-type BRCA1 cDNA, wherein a difference in the sequence of the BRCA1 gene, BRCA1 RNA, or BRCA1 cDNA of the subject from wild-type indicates an alteration in the BRCA I gene in said subject.[5]

As this patent claim highlights, disease gene patents basically lay claim to an act of observing DNA in the form in which it exists in individuals' cells when such observing is performed for the purpose of diagnosis. The act of looking itself is not new. Many techniques such as Southern analysis, sequencing, PCR, and others may be used by a geneticist to determine the genetic makeup of an individual at a specific locus. The sequences of the genes themselves may not even be new because the sequence may be in a public database somewhere. What is new is the empirically discovered association between different isoforms of a gene and some phenotype, be it eye color or susceptibility to disease. That is, what is new is knowing where to look.

While the tools for looking and the methods for using those tools (e.g., PCR) may be patentable, we do not believe that any one specific act of looking should be patentable because everyone should be free to look at the world around us without restraint. This self-evident truth we do not believe to be altered by either the difficulty of looking (be it with the naked eye, a microscope, or gene sequencing) or the market value of the view. The issue as we see it is whether one can patent the act of observing something as it exists in its "natural state" merely because that observation is found to have utility (beyond the pleasure of the experience). One may not patent the "natural" thing itself, so why should the act of observing the thing be patentable?

There is room for disagreement on the answer to this question. Indeed, a Lexis patent search revealed at least one other instance where specific acts of observing have been patented: for the detection of mineral deposits or oil reservoirs for the purpose of determining where to drill.[6] For example, U.S. Patent No. 3,919,547 claims a method of estimating the distal expanse of an oil reservoir by measuring gamma radiation emitted by two different naturally occurring isotopes in different locations on the surface near an oil field where a change in the ratio of the field strength indicates the bounds of the reservoir. U.S. Patent No. 4,939,460 claims the surface measurement of Earth's geomagnetic field in order to detect a subterranean geologic deposit. Similarly, U.S. Patent No. 4,053,772 claims the use of dosimeters to measure alpha particles from radon decay in order to detect underground uranium deposits. These patents show that the acts of observing (at least when using a tool other than the naked eye and performed for a useful purpose) indeed comprise patentable subject matter. However, disease gene patents may still be distinguished from these oil hunting claims because each disease gene patent is drawn to use by any of the known tools of observation for examination of one narrowly defined chemical structure. Instead of claiming the use of method X for determining whether oil exists underground at any location, disease gene patents are analogous to claiming the use of any method for determining whether oil exists in, say, Texas.

Nonetheless, appeals to metaphysics as voiced above cannot provide a definitive answer for what is "natural." As Daniel Kevles notes (this volume), what is patentable is ultimately a political decision. Indeed, as the pragmatist Glenn McGee argues, patentability perhaps should be determined by whether there is a market.[7] My colleagues and I have thus turned our attention from whether disease gene patents should be permitted to examining how they are, in fact, being used.

What we have found is that these patents are often being exclusively licensed to for-profit laboratories by academic medical centers where the basic biomedical science was performed.[8] The exclusive arrangements allow patent holders to enforce their patents in ways that infringe on the practice of medicine, restrict clinical observation and formal research, reduce access, and increase the costs of clinical testing services.[9] One ongoing case in which I have been involved presents an extreme example of the mistreatment of human research subjects and corporate greed. I use the case here to flesh out questions about who rightfully can claim ownership in the products of genomic discovery.

THE CANAVAN DISEASE PATENT CASE

The Research

In 1981 Dan and Debbie Greenberg had a son, Jonathan. After Jonathan failed to thrive, he was diagnosed with Canavan disease. Canavan is a recessive genetic disease that will strike on average one of four children of couples when both parents carry a gene mutation that causes the disease. It is a degenerative spongiform brain disease that irreversibly leads to loss of body control and death, usually before the teen years.[10] There is no cure.

In 1983 the Greenbergs had another child, Amy, who, against the odds, also was stricken with the disease. In 1987 Dan Greenberg approached Dr. Reuben Matalon and convinced him to study Canavan disease. Matalon ran a laboratory performing clinical testing and research of phenylketonuria (PKU) and other familial disorders at the University of Illinois in Chicago. With blood, urine, other tissue samples provided by the Greenbergs and another family affected by the disease, and "seed money" provided by Greenberg's Chicago chapter of the National Tay-Sachs and Allied Diseases Association (NTSAD), within a year Matalon identified the deficiency of an enzyme, aspartoacylase, as the cause of Canavan disease. This was great news because it offered the possibility of a prenatal screening test.[11]

In 1988 the Greenbergs became the first couple to commence a pregnancy based upon knowledge of the availability of the testing. They underwent prenatal testing with Matalon's enzymatic assay and gave birth to a healthy child. This was repeated for at least nineteen other couples over the next two years.[12] During this time, Matalon, joined by his colleague Rajinder Kaul, moved to the Miami Children's Hospital (MCH) Research Institute. It was at the MCH, in the early 1990s, that Matalon's laboratory misdiagnosed four pregnancies that resulted in the birth of children with Canavan disease. The misdiagnoses also resulted in at least two lawsuits, which were settled out of court.[13] Matalon and Kaul had begun looking for the gene, which offered the only reliable method for prenatal testing as well as carrier screening.

Guangping Gao, then a graduate student at Florida International University working in the lab of and under the tutelage of Dr. Kaul, succeeded in cloning the Canavan gene by early 1993.[14] The research drew on tissue samples provided to Matalon by the Greenbergs and over one hundred other families from around the world who had been stricken by the disease as well as blood samples provided by Rabbi Josef Ekstein, executive director of Dor Yeshorim Committee for the Prevention of Jewish Genetic Diseases in Brooklyn, New York. Rabbi Ekstein provided about six thousand stored blood samples that were used by the researchers to rapidly identify several mutations in Ashkenazi Jewish families and estimate population frequencies.[15]

The Patent and Licensing

A patent application was filed in September 1994, and U.S. Patent No. 5,679,635 was issued to the MCH Research Institute in October 1997. It is to date the only patent owned by the hospital or its research institute.[16] The MCH began working with Marc Golden, an intellectual property consultant from New York, on a marketing plan for its patent. At the same time, advocates from the Canavan Foundation in New York and the NTSAD in Boston were working with local and national groups to promote Canavan disease testing and were successful in convincing the American College of Obstetricians and Gynecologists to issue guidelines recommending carrier screening of Ashkenazi couples.[17] Within two weeks after those guidelines were released, the MCH began sending letters to clinical laboratories informing them of the patent and the hospital's plans for commercializing the test. Rabbi Ekstein received such a letter at Dor Yeshorim.[18]

The marketing plan consisted of two stages of licensing. In the first stage, a limited number of academic laboratories (likely to be a subset of the many already performing the testing) would be granted nonexclusive licenses to perform an annually limited number of tests. A fixed $12.50 per test royalty was demanded, and according to Golden, a number of laboratories signed license agreements.[19] In the second stage, a large commercial laboratory would be licensed as a "market leader" with what would be, in effect, an exclusive license to the remainder of the testing volume. The justification for this plan was that a large reference laboratory would be able to spend the resources for outreach and education needed to ensure screening and testing of all couples at risk. What this plan ignored was the role of community organizations—such as NTSAD, local temple and consumer groups, and Dor Yeshorim—that were instrumental in making Tay-Sachs carrier screening widely available and used, and the United Leukodystrophy Foundation, which had developed a registry and screening program for Canavan disease promptly after the gene was discovered. Tay-Sachs testing methods have never been restricted, and NTSAD and Dor Yeshorim stand as testaments to the ability of community-based organizations to develop and carry out population education and screening. In lieu of engaging these groups, the MCH and Golden sent letters indicating their intent to aggressively enforce the patent.

In 1999 the Canavan Foundation, the NTSAD, the National Foundation for Jewish Genetic Diseases, and the Canavan Research Fund created the Canavan Disease Screening Consortium. In July 1999 the consortium attempted to contact and establish a dialog with the MCH, but was unsuccessful. On November 14, 1999, an article describing the Canavan patent issue was published in the *Miami Herald*.[20] The consortium followed up the article with several advertisements criticizing the MCH for its licensing position, and several additional newspaper articles appeared discussing the case.[21] Finally, they received a letter from the president of the MCH, Thomas Rozek, welcoming the opportunity to talk. A meeting was scheduled for January 20, 2000.

The consortium secured the participation of two experts to help them make the case to MCH management that the proposed licensing arrangement and royalty is harmful. Dr. Michael Watson, a geneticist from Washington University of St. Louis, was invited because of his clinical expertise, and I was invited to discuss ethical concerns about restrictive licensing. When the MCH learned that I had collaborated with Dr. Debra Leonard, director of the Molecular Pathology Laboratory at the University of Pennsylvania Health System, with whom they were involved in negotiations for one of the stage I licenses, they requested that I not be included.[22] The hospital also began negotiating a set of rules for the meeting, which most significantly imposed a complete and permanent gag on all participants. When the MCH found out that the anticipated meeting had been mentioned in an article in the *Boston Globe* that appeared on December 20, 1999, they almost cancelled the meeting.[23] Finally, MCH management agreed that the consortium members, including the two experts, could come and make a presentation. The MCH promised only that they would enter no new agreements for seven days after the meeting, and that they would provide a response to the consortium in not less than thirty days. The consortium promised not to initiate any public criticisms or advertising programs for thirty days.

Following the meeting, the MCH, through its agent Golden, continued license negotiations with Dr. Leonard. The fundamental conditions of the license agreement were (1) $12.50 fixed royalty per test, (2) the annual number of tests Leonard's laboratory could perform would be limited to 150 percent of the volume in the first year of the license, (3) that the license could be terminated by the MCH at will, and (4) that Leonard could say nothing about the license except that it was limited and nonexclusive. The confidentiality clause would survive termination of the license for any reason. During negotiations, Golden remarked to university counsel that Dr. Leonard had been quite outspoken on the issue of gene patenting, and the MCH would not remove the clause of the draft license that would enable the hospital to terminate it at will. Dr. Leonard has written several articles and given numerous talks criticizing disease gene patents, and she is current president of the Association for Molecular Pathology, which has adopted a formal position statement denouncing exclusionary licensing of clinical testing patents.[24] Dr. Leonard perceived the license as a substantial infringement on her academic and medical freedom, as well as a dangerous impediment to her ability to deliver quality laboratory services. On advice from counsel, she refused to agree to the terms offered.

This was not the end of the negotiation. Dr. Leonard's laboratory had performed Canavan disease testing since the patent had been issued, and the MCH wanted their royalty payment for those tests. Golden drafted a settlement agreement for the previously performed tests and included a term that provided that no physicians at the University of Pennsylvania could "perform, or have other(s) perform, any Canavan Tests . . . without first obtaining a license," which Dr. Leonard had been counseled not to sign. This was not a typographical error, and Mr. Golden confirmed to university counsel that the intent was to prohibit all physicians at the University of Pennsylvania from sending samples to any licensed laboratory for Canavan disease testing unless a license was taken. This can only be interpreted to be a transparent act of retribution for the university's refusal to accept the MCH's terms, and it defies all conceptions of fairness, justice, and common decency. Apparently, the MCH wanted university physicians to cease the practice of medicine, mindless of the human costs in missed opportunities to screen couples and avoid the devastation of Canavan disease. The university, of course, refused to execute such an agreement and finally agreed to pay the past royalty and simply warrant that they would not infringe on the patent in the future.

It does not end there, either. When the license was refused, Dr. Leonard requested, through university counsel, information on the laboratories that held licenses so that she could send samples to a licensed laboratory for testing. The MCH stated that it would identify only four licensed laboratories out of about a dozen that had taken licenses. Concurrently, the consortium, led by Orren Alperstein Gelblum of the Canavan Foundation and Judith Tsipis of NTSAD, was attempting to learn from the MCH what labs held licenses. These groups were actively engaged in educating the community at risk and needed to know where to refer couples for testing. The MCH refused to provide this information. Nor did they offer any reasons why, if they were so interested in ensuring that testing be promoted, they would not disclose this information to those in a position to educate the at-risk community about where the licensed testing services could be found. It may be guessed that the hospital was attempting to find its "market leader" laboratory, and was holding open the option of granting one exclusive license and terminating the limited licenses it had already granted to academic laboratories. If that happened, then all testing "business" would be referred to the exclusive licensee.

In early April 2000, the MCH gave up its attempt to find a "market leader" laboratory. On April 3, 2000, the MCH mailed a letter to the consortium offering about $20,000 per annum of an estimated $375,000 in royalties, to be used to increase public awareness and to help provide testing to those unable to afford it.[25] The consortium rightly believed that if they were unable to dissuade the MCH from collecting royalties on the test, then the hospital should dedicate some of the revenue to increasing awareness and access by the at-risk population. The offer, however, carried the condition that consortium members no longer be publicly critical of the MCH regarding the Canavan disease gene patent. In response, the consortium welcomed the financial help with their outreach programs, but refused to agree to no longer be free to speak about the hospital's licensing program and royalty fees, issues with which they fundamentally disagree.

Ethical Issues

There are numerous ethical issues raised by this case. Many of the general issues have been discussed elsewhere, and thus I discuss here the three most salient issues.[26]

Business Secrecy

One issue that jumps off the page is the secrecy sought by the MCH at all stages of its dealing. Secrecy is a fundamental value in business, but it strikes against the values of public health and medicine.[27] The MCH sought to hide everything behind a cloak of secrecy, from the meeting they agreed to hold with Canavan activist families to the terms of licenses with academic laboratories. They have even kept secret the identities of laboratories they have licensed from those who most need the information.

Secrecy in licensing of disease gene patents appears to be common. According to the MCH, about a dozen laboratories agreed to their licensing terms, which presumably contain strict confidentiality clauses. In an attempt to understand how disease gene patents are being licensed, I have sought copies of several license agreements covering patents resulting in part from federally funded research. In November 1998 I asked the National Institute of Health Office of Technology Transfer and the vice president of research at the University of Utah for copies of the licenses negotiated between those institutions and Myriad Genetics for U.S. Patent No. 5,753,441, covering the BRCA1 gene and tests. Both institutions reported that such licenses are considered proprietary, and thus are not subject to Freedom of Information Act (FOIA) requests. In October 1998 I filed a state FOIA request for the exclusive license from the University of Minnesota to Athena Diagnostics for U.S. Patent No. 5,741,645, which covers genetic testing for spinocerebellar ataxia type 1. The agreement reportedly was sent to Athena for redaction of proprietary information, and, after a year and a half, the license has yet to be produced.

Secrecy seems to have taken on some fundamental value, rather than being merely instrumental in the protection of trade information that could be competitively harmful to the organization. Because of such secrecy, there simply is little known about how the licensing of genetic technology may be influencing the dissemination and use of tests, including whether licensors are demanding "reach-through" rights to the products of subsequent research and development. As the Canavan disease case exposes, licensing practices may impose serious costs and limitations on the availability of laboratory services.

Ethics of Research

According to some of the involved families, no consent form was used for much of the early research on Canavan disease. This was in violation of ethical norms.[28] Dr. Matalon simply sent out Guthrie cards (filter papers used to preserve blood for genetic analysis) and collected blood, names, and social security numbers from affected children and parents. The families most directly involved helped by identifying, contacting, informing, and soliciting participation of other affected families. It was these families who, in 1994, suggested to Matalon that a consent form should be used, and who also helped generate a form that was used thereafter. According to Matalon, all his research was approved by the MCH's institutional review board, presumably with an express waiver of informed consent.[29] It certainly does not appear to me that there is any justification for a waiver of fully informed consent to the prospective collection of blood and other tissue samples for genetic studies.

Furthermore, no participants ever suspected that the discovery of "their" gene could, or would, result in a patent.[30] Thus, these families—who had participated in a very active way in the research enterprise of Reuben Matalon in hopes of helping families like them avoid the ravages of Canavan disease—were dumbfounded when the MCH was issued a U.S. patent covering the genetic test for the disease, and began to enforce it. Not only were they uninformed about the scope of their research participation, but they were betrayed by the ultimate commercialization and profiteering of the institution they believed was motivated, as were they, by altruism and the desire to help prevent this terrible disease. As my colleague Jason Karlawish has suggested, participants should have a say about the nature of the benefits that may result from the research, particularly acute when a community is involved closely in the performance of the research.[31] In this case, there clearly was a mismatch between what participants and the community expected and what the researcher and his institution sought and achieved.

Ownership of Genetic Invention

There were three necessary ingredients in the discovery of the Canavan gene. First was the active pursuit of research and willing study participation by families stricken by the disease. These parents were highly motivated by their desire to avoid births of children with Canavan disease in order to minimize the pain to families and to their community from this disease. They were willing to be research subjects so that medical science could be advanced, and their participation was a commonly accepted altruistic gift that would hopefully benefit them as well as others. Second was the intellectual capital invested both by the researchers who worked on this particular research and many, many other scientists who contributed to the great advances in genetic technologies over the last two decades. These scientific advances were used effectively by Drs. Matalon, Kaul, and Gao in working from an identified enzyme to cloning the target gene. These individuals trained themselves to perform genetics research, and they performed wonderfully. Third was the financial capital that was provided to these scientists to perform the research. Research funding totaling an estimated $5 million over several years was provided by numerous donors to the MCH Research Institute, including the NTSAD, the Canavan Foundation, various local Jewish organizations, and the United Leukodystrophy Foundation.

Interestingly, the patent only rewards those who have made financial or intellectual investments, not those who provided what Matalon (and other researchers) called "human resources."[32] Nonetheless, the families played what they believe to be significant roles in the research leading to the discovery of the Canavan disease gene. Dan Greenberg stated that the gene discovery was the result of an "extraordinarily productive partnership between lay parents and a medical researcher," and he feels a paternal role in the research, which he describes as being the "grandfather" of the discovery. Reflecting a starkly contrasting view, Reuben Matalon told me that Dan Greenberg played "no role" in the discovery of the gene, despite contributing samples, helping identify and solicit families to participate in the research, and providing the aforementioned "seed money" for the early work.[33] Matalon's view has apparently mutated over time, because he was quoted in 1993 as saying, "This is a disease where a partnership between researchers and the families of affected children is critical for advancing knowledge for prevention, and, hopefully, for helping affected children."[34]

As mimicked in the comic here, and drawing on the observation of Renee Fox and Judith Swazey in the sociology of transplantation, researchers may be characterized as being akin to predators in their approach to the persons who provide "resources" for their studies.[35] If sub-

"There. Now it's all on paper. Feel better?"

jects are actively misled or even passively permitted by researchers and
institutional review boards to proceed under false impressions and beliefs
about issues in the research that could be material to their willingness to
participate, then their consent is absolutely meaningless. Fima Lifshitz,
chief of the medical staff at the MCH, was quoted in the *Miami Herald* as
saying, "You voluntarily submitted a blood sample to be tested. As a result,
I discover a gene that's patentable. What's wrong with that? This is done all
the time. The issue should be quenched at once because these people are
going to derive a great deal of benefit from this. They shouldn't be com-
plaining." The problem with Dr. Lifshitz's analysis, in part, is that subjects
were not submitting samples to be tested, but were participants in an elaborate research scheme. More problematic is the assertion of voluntariness
when there was no informed consent. There was such a complete mismatch between the expectations and goals of subjects and researchers.

The contributions of the researchers and of the financial donors who
supported the research can be said to be fungible; the only irreplaceable,
"critical" resource—the *sine qua non*—in the discovery of the gene was the
participation of the affected families. The help of Rabbi Ekstein in providing thousands of samples from his clinical collections was, simply, irreplaceable. Thus, it is extremely ironic that the system, in the end, fails to
acknowledge their status and their contribution. It is, in a very real way,

"their" gene; yet, not. The ambiguous status of genes is reflected perhaps best in the quote of Guangping Gao, who performed the bulk of the laboratory research leading to the discovery of the gene. At a press conference announcing the discovery, upon being introduced to Jacob Eisen, one of the subjects in the research, Gao said to Jacob's mother: "I cloned his gene. I held his gene in my hand. It's nice to meet him."[36]

DISCUSSION

The Canavan disease gene case is perhaps an extreme example of what can go wrong with patents when commercial interests infect medicine. The scandalous nature of the behaviors described may be attributed to corporate greed and the subversion of public and medical welfare to parochial managerial interests in the bottom line. Simply, the corporate decision makers have a patent that is worth money, and they are in no way held accountable to broader obligations to do good for society, for the at-risk population, for those who contributed money for the support of this research effort, and for those who participated in the basic biomedical research that yielded the patentable discovery.

Of course, patents play an extremely useful role in providing incentives and rewards for pharmaceutical development. Patents are also viewed—at least by the stock market—as a very important and valuable system for the biotechnology industry, which holds the promise of large payoffs in the future of gene-based therapies and cures. But patents are also viewed with skepticism when they are used to tie up fundamental knowledge that is many steps removed from a therapeutic agent, where monopoly and the "anti-commons" described by Heller and Eisenberg may create substantial disincentives to research and development.[37] As described in the Canavan case, the exclusionary licensing of gene patents may also infringe on the practice of medicine.

This case also highlights what seems to be a fundamental unfairness in the definition and distribution of benefits from research. It is reasonable to ask what motivated each party to contribute to the research, what their expectations were regarding what benefits were likely to result from the research and to whom those benefits would accrue, and whether the financial rewards that resulted enured to the various involved parties in an equitable manner.

More important, the Canavan case suggests that patient groups must adopt a more aggressive role in the performance of research. Very recently, the gene that causes pseudoxanthoma elasticum (PXE), a genetic disease of the connective tissue, was found in research that was pushed forward by the PXE International foundation.[38] The foundation helped identify and solicit participation of affected families, set up a repository, and raised money to support study. It also negotiated with researchers to whom it gave support and materials for research and retained rights in any patents in order to ensure broad and affordable availability of the test and any downstream developments.[39] What is most informative about the PXE counterexample is that it sheds light on the inability of institutional review boards to adequately protect all of the interests of potential subjects in research.

The law provides no after-the-fact remedies for patients and research subjects who may come to believe they have been taken advantage of.[40] While there have been various recommendations that shares of profits flowing from research be provided to subjects or their communities, no formal laws, regulations, or policies within the research community have been promulgated to date to make this happen.[41] While it may be difficult for subjects, patient groups, researchers, and those who sponsor research (such as biotechnology and pharmaceutical companies) to come to agreement about how the benefits of research should be shared, it seems quite clear that these issues should be openly discussed and resolved before research is done. The extra mile should be traveled by researchers and, perhaps even more important, by institutional review boards to engage patient groups and others from relevant, affected communities in decisions about what will happen with the fruits of research.

ACKNOWLEDGMENT

Supported in part by grants from the Greenwall Foundation, the Charles E. Culpeper Foundation, and the NIH, DOE, and VA Consortium on Informed Consent Research, and by a grant from the Dodge Foundation. The opinions expressed are those solely of the author. The author especially thanks Judith Tsipis, Daniel Greenberg, and Orren Alperstein Gelblum for sharing their personal stories and experiences with Canavan disease and the described research project, for actively involving the author in their unfolding saga with Miami Children's Hospital, for freely sharing background information on the case, and for reading and commenting on earlier drafts of this manuscript; knowledge about the Canavan disease gene, like this chapter, would not exist but for their contributions. The author also thanks Mildred K. Cho and Debra G. B. Leonard for continued collaboration on gene patenting issues and for their comments on this manuscript; Reuben Matalon, Rajinder Kaul, Guangping Gao, Ben Roa, and Rabbi Ekstein for sharing their experiences with Canavan disease; and Rajinder Kaul, Marcie Merz, and Arthur Caplan for comments on earlier versions of the manuscript.

NOTES

1. House Rept. 1923, 82d Cong., 2d Sess. 6, 1952; Senate Rept. 1979, 82d Cong., 2d Sess. 5, 1952.

2. *Diamond* v. *Chakrabarty*, 447 U.S. 303, 308–9 (1980).

3. *In re Bergy*, 596 F 2d 952, 988 (CCPA 1979) (Baldwin, concurring).

4. Jon F. Merz, Mildred K. Cho, Madeline J. Robertson, and Debra G. B. Leonard, "Disease Gene Patenting Is a Bad Innovation," *Molecular Diagnosis* 2 (1997): 299–304.

5. U.S. Patent No. 5,753,441.

6. This was suggested by several patent lawyers who responded to an unpublished Internet survey. Interestingly, the lawyers disagreed about whether the methods of observing and predicting the presence of oil would comprise patentable subject matter. The results of this survey are available from the author upon request.

7. Glenn McGee, "Gene Patents Can Be Ethical," *Cambridge Quarterly* 7 (1998): 417–21.

8. Anna Schissel, Jon F. Merz, and Mildred K. Cho, "Survey Confirms Fears about Licensing of Genetic Tests," *Nature* 402 (1999): 118.

9. Mildred K. Cho, "Ethical and Legal Issues in the Twenty-First Century," in *Preparing for the Millenium: Laboratory Medicine in the 21st Century*, 2d ed., (Washington, D.C.: American Association for Clinical Chemistry Press, 1998); Jon F. Merz, "Disease Gene Patents: Overcoming Unethical Constraints on Clinical Laboratory Medicine," *Clinical Chemistry* 45 (1999): 324–34.

10. Reuben Matalon, Kimberlee Michals, and Rajinder Kaul, "Canavan Disease: From Spongy Degeneration to Molecular Analysis," *Journal of Pediatrics* 127 (1995): 511–17.

11. Reuben Matalon, Kimberlee Michals, Dana Sebesta, Minerva Deanchng,

Peter Gashkoff, and Jose Casanova, "Aspartoacylase Deficiency and N-acetylas-partic Aciduria in Patients with Canavan Disease," *American Journal of Medical Genetics* 29 (1998): 463–71.

12. Reuben Matalon, Kimberlee Michals, Peter Gashkoff, and Rajinder Kaul, "Prenatal Diagnosis of Canavan Disease," *Journal of Inherited Metabolic Diseases* 15 (1992): 392–94.

13. Michael Winerip, "Fighting for Jacob," *New York Times Magazine*, 27 December 1998.

14. Rajinder Kaul, Guangping Gao, Kuppareddi Balamurugan, and Reuben Matalon, "Cloning of the Human Aspartoacylase cDNA and a Common Missense Mutation in Canavan Disease," *Natural Genetics* 5 (1993): 118–23.

15. Rajinder Kaul, Guangping Gao, Maria Aloya, Kuppareddi Balamurugan, Arlene Petrosky, Kimberlee Michals, and Reuben Matalon, "Canavan Disease: Mutations among Jewish and Non-Jewish Patients," *American Journal of Human Genetics* 55 (1994): 34–41.

16. June 21, 2000, search of the IBM patent database, <http://patent.wom-plex.ibm.com/>. Interestingly, the patent lists as "inventors" people other than Gao and Kaul, including Matalon's wife, Kimberlee Michals-Matalon. According to Dr. Kaul, Michals-Matalon had "absolutely nothing to do with the research" performed in his laboratory, and she does not appear as an author of the Nature Genetics paper, supra note 14. Matalon also stated that he "added" Kaul, Gao, and others to the patent, "because that is what is done." Actually, inventorship is strictly defined by the patent statute, U.S. Code Title 35, as one who makes a contribu-tion to the subject matter of the claims of a patent. Falsely adding or excluding inventors can result in patent invalidity, and this practice appears to be all too common in biotechnology. See Philippe Ducor, "Coauthorship and Coinventor-ship," *Science* 289 (2000): 873–75.

17. American College of Obstetricians and Gynecologists, Committee on Genetics, "ACOG Committee Opinion: Screening for Canavan Disease," No. 212, November 1998, *International Journal of Gynaecology & Obstetrics* 65 (1999): 91–92.

18. Interview with Rabbi Josef Ekstein, April 2000. The licensing arrange-ment was in part responsible for the breakdown of the testing arrangement that Dor Yeshorim had with Baylor University. Interview with Dr. Benjamin Roa, May 2000. Ekstein stated that there is "no question [MCH's licensing program] is going to bring more Canavan's children into the world."

19. Karen Rafinski, "Hospital's Patent Stokes Debate on Human Genes," *Miami Herald*, 14 November 1999, 1.

20. Ibid.

21. See, for example, Judith Peres, "Genetic Tests Reduce Neighborhood's Grief: Screening Stops Unwise Matches," *Chicago Tribune*, 12 September 1999, 16;

"Flap Erupting over Royalty for Canavan: Miami Children's Hospital Exercises Patent for Test," *Forward* (New York), 20 August 1999, 15.

22. It is interesting to note that in Rozek's November 22 letter to Orren Alperstein Gelblum of the Canavan Foundation, he states that MCH had "just reached agreement with the last of the mid-sized laboratories that were targeted in the first phase of our licensing program. All of these laboratories have agreed to the royalty fee structure proposed." Yet, a mere six days before, MCH had sent a letter to Debra Leonard compelling her to negotiate a license under the patent.

23. Richard Saltus, "Critics Claim Patents Stifle Gene Testing," *Boston Globe*, 20 December 1999.

24. An early version of this policy statement was also adopted by the Academy of Clinical Laboratory Physicians and Scientists, <http://www.pathology.med.umich.edu/aclpsnewJul99/license.htm>, (9 June 2000). The statement is currently under negotiation and revision by a large number of pathology and genetics organizations. See, for example, College of American Pathologists, *Genes Patents Detrimental to Care, Training, Research.* <http://www.cap.org/html/Advocacy/issues/genetalk.html>, (14 July 2000).

25. Rafinsky, "Hospital Patent Stokes Debate on Human Genes."

26. One of the concerns we have expressed is that any restriction on the delivery of clinical testing services may reduce widespread clinical observation, validation, and advances in medical training and knowledge. Indeed, many new mutations have been discovered since the original paper was published and patent filed. See Peter L. Rady, Joseph M. Penzien, Trini Vargas, Stephen K. Tyring, and Reuben Matalon, "Novel Splice Site Mutation of Aspartoacylase Gene in a Turkish Patient with Canavan Disease," *European Journal of Pediatric Neurology* 4 (2000): 27–30; Peter L. Rady, Trini Vargas, Stephen K. Tyring, Reuban Matalon, and Ulrich Langenbeck, "Novel Missense Mutation (Y231C) in a Turkish Patient with Canavan Disease," *American Journal of Medical Genetics* 87 (1999): 273–75; Reuben Matalon, Maria Aloya, Quin Su, Mei Jin, Anne B. Johnson, Ruud B. Schutgens, and Joe T. Clark, "Identification and Expression of Eight Novel Mutations Among Non-Jewish Patients with Canavan Disease," *American Journal of Human Genetics* 59 (1996): 95–102; and Rajinder Kaul, Guangping Gao, Kimberlee Michals, D. T. Whelan, S. Levin, and Reuben Matalon, "Novel (cys152>arg) Missense Mutation in an Arab Patient with Canavan Disease," *Human Mutations* 5 (1995): 269–71.

27. Steven A. Rosenberg, "Secrecy in Medical Research," *New England Journal of Medicine* 334 (1996): 392–94.

28. Interviews with Dr. Judith Tsipis, Ms. Orren Alperstein Gelblum, and Mr. Daniel Greenberg, January 2000.

29. According to Rajinder Kaul, several proposals for NIH funding of the

research were submitted, but were not funded. IRB approval and compliance with the federal Common Rule, 45 C.F.R. Part 46, would have been strictly required before such funding was provided.

30. Indeed, in an interview, Reuben Matalon stated that he did not contemplate patent protection until he spoke with a colleague from Scripps Research Institute, Ernest Beutler, who had several patents on mutations associated with Gaucher disease. See U.S. Patent Nos. 5,234,811 and 5,266,459. Scripps Research Institute is granting nonexclusive licenses to these patents, with a royalty of $5 per test. Phone interview with Donna Weston, 14 July 2000.

31. Jason Karlawish and John Lantos "Community Equipoise and the Architecture of Clinical Research," *Cambridge Quarterly Healthcare Ethics* 6 (1997): 385–96.

32. This is from Matalon's acknowledgement in the original enzyme paper. *American Journal of Medical Genetics* 29 (1988): 469–70.

33. Interviews with Drs. Reuben Matalon, Guangping Gao, and Rajinder Kaul, April 2000.

34. United Leukodystrophy Foundation, press release, "Canavan Disease Carrier Screening Program Established by the United Leukodystrophy Foundation." Sycamore, Ill., 28 September 1993.

35. Renee Fox and Judith Swazey, *Spare Parts* (New York: Oxford University Press, 1992); see also Lori Andrews and Dorothy Nelkin, "Whose Body Is It Anyway? Disputes over Body Tissue in a Biotechnology Age," *Lancet* 351 (1998): 53–57.

36. Elinor Brecher, "S Florida Research Offers New Insight into Rare Disease," *Miami Herald*, 29 September 1993, 1A, 21A.

37. "Patent Rules Should Include a Defence against Monopolies," *Nature* 399 (1999): 619; Michael Heller and Rebecca Eisenberg, "Can Patents Deter Innovation? The Anticommons in Biomedical Research," *Science* 280 (1998): 698–701.

38. PXE International <http://www.pxe.org/>, (8 June 2000).

39. Gina Kolata, "Who Owns Your Genes?" *New York Times*, 15 May 2000, A1.

40. *Moore v. Regents of University of California*, 51 Cal 3d 120, 271 Cal Rptr. 146, 793 P 2d 479 (1990).

41. See, for example, Robert Weir and Jay Horton, "DNA Banking and Informed Consent—Part 2," *IRB: A Review of Human Subjects Research* 17, nos. 5–6 (1995): 1–8 (suggesting that as much as 10 to 25 percent of profits be returned to subjects); and, HUGO Ethics Committee, "Statement on Benefit-Sharing," London: Human Genome Organization, 9 April 2000 (suggesting that profit-making entities dedicate 1 to 3 percent of profits to healthcare infrastructure or humanitarian efforts); and Merz et al., "Protecting Subjects' Interests in Genetic Research," *American Journal of Human Genetics* 70 (2002): 965–71.

SIX

HOW CAN YOU PATENT GENES?
REBECCA S. EISENBERG

A s rivalrous initiatives in the public and private sectors raced to complete the sequence of the human genome,[1] patent issues played a
prominent role in speculations about the significance of this achievement.[2]
How much of the genome would be subject to the control of patent
holders, and what would this mean for future research and the development
of products for the improvement of human health?[3] Is a patent system
developed to establish rights in mechanical inventions of an earlier era up to
the task of resolving competing claims to the genome on behalf of the many
sequential innovators who elucidate its sequence and function, with due
regard to the interests of the scientific community and the broader public?[4]

A deeper question is logically prior to these more fine-grained
inquiries: How can you patent DNA sequences? Indeed, over the course
of fifteen years of giving talks on the topic of biotechnology patents to
widely varying audiences, this has been the question that I am asked most
frequently and persistently. Although patent applicants have been seeking
and obtaining patent claims on DNA sequences for some twenty years
already,[5] many people find this practice troubling and counterintuitive.
One might expect that the U.S. Patent and Trademark Office (PTO) and
the courts would have resolved this issue many times over as the industry
pursued and litigated patent claims covering biotechnology products, and
that biotechnology patent law would now be entering a relatively mature
phase in which fundamental questions have been resolved and the issues
that remain to be addressed are incremental and interstitial.[6] Instead, the
patent system is struggling to clarify the ground rules for patenting DNA
sequences, while years' worth of patent applications accumulate in the

PTO. What accounts for this persistent lack of clarity about how patent law applies to this technology?

A significant part of the problem lies in the shifting landscape of discovery in genetics and genomics research. The patent system, which inevitably requires time to resolve even routine matters, has so far focused primarily on the discoveries of the 1980s.[7] The DNA sequences that were the subject of patent claims in this era typically consisted of cloned genes that enabled the production of proteins through recombinant DNA technology. Patents on the genes and recombinant materials that supply the genetic blueprint for these proteins promised exclusivity in the market for the protein itself, equivalent to the protection that a pharmaceutical firm obtains by patenting a new drug. From this perspective, patents on DNA sequences seemed analogous to patents on new chemical entities. The Court of Appeals for the Federal Circuit accordingly turned to prior cases concerning patents on chemicals in resolving disputed issues about how patent law should apply to DNA sequences.[8] Whatever the limitations of this analogy, it provided a relatively clear point of departure for analyzing legal issues presented by patent claims to the first generation of biotechnology products that came to market—therapeutic proteins produced through recombinant DNA technology.

As DNA sequence discovery has moved beyond targeted efforts to clone particular genes to large-scale, high-throughput sequencing of entire genomes, new questions have come into view. The DNA sequences identified by high-throughput sequencing look less like new chemical entities than they do like new scientific information. From the perspective of patent claimants, the chemical analogy is of little value as a strategic guide to capturing the value of this information as intellectual property. From the perspective of the PTO and the courts, claims to these discoveries raise unresolved issues on which the chemical analogy sheds little light. The result is profound uncertainty concerning the meaning of the doctrinal tools that the patent system offers for determining what may be patented and for drawing boundaries between the rights of inventors and the rights of the public.

PATENT ELIGIBILITY

A threshold issue that one might expect to have been resolved long ago is whether DNA sequences are the sort of subject matter that the patent system protects. The U.S. patent statute defines patent-eligible subject matter as "any process, machine, manufacture, or composition of matter,"[9] language that the U.S. Supreme Court has held indicates an expansive scope that includes "anything under the sun that is made by man."[10] Although cases have held that "products of nature" may not be patented, this exclusion has not presented an obstacle to obtaining patents claiming DNA sequences in forms that do not occur in nature as new "compositions of matter." On the threshold issue of patent-eligible subject matter, as on other issues, the analogy to chemical patent practice has supplied an answer.

The standard patent lawyer's response to the "products of nature" intuition is to treat it as a technical, claim-drafting problem. From this perspective the prohibition against patenting products of nature only prevents the patenting of DNA sequences in a naturally occurring form that requires no human intervention. One cannot get a patent on a DNA sequence that would be infringed by someone who lives in a state of nature on Walden Pond, whose DNA continues to do the same things it has done for generations in nature. But one can get a patent on DNA sequences in forms that only exist through the intervention of modern biotechnology.

Patents have thus issued on "isolated and purified" DNA sequences, separate from the chromosomes in which they occur in nature, or on DNA sequences that have been spliced into recombinant vectors or introduced into recombinant cells of a sort that do not exist on Walden Pond.[11] This is consistent with the long-standing practice, even prior to the advent of modern biotechnology, of allowing patents to issue on isolated and purified chemical products that exist in nature only in an impure state, when human intervention has made them available in a new form that meets human purposes.[12] This is not simply a lawyer's trick, but a persuasive response to the intuition that patents should only issue for human inventions. It prevents the issuance of patents that take away from the public things that they were previously using (such as the DNA that resides in their cells), while allowing patents to issue on new human manipulations of nature. Those of us who simply use the DNA in our own cells, as our ancestors have been doing for

generations, should not and need not worry about patent infringement liability. On the other hand, those of us who get shots of recombinant insulin
or erythropoietin can in fairness expect to pay a premium to the inventors
who made these technological interventions possible.

MOLECULES VERSUS INFORMATION

The patentability of DNA molecules in forms that involve human intervention appears to be well settled. But recent advances in DNA sequencing
present the patent eligibility issue from a somewhat different angle that the
courts have yet to address. DNA sequences are not simply molecules, they
are also information. Knowing the DNA sequence for the genome of an
organism provides valuable scientific information that can open the door to
future discoveries. Can the value of this information be captured through
patents? Can information about the natural world, as distinguished from tangible human interventions that make use of that information, be patented?

The traditional statutory categories of patent-eligible subject matter—
processes, machines, manufactures, and compositions of matter—seem to
be limited to tangible products and processes, as distinguished from information as such. Although many cases have used the word "tangible" in
defining the boundaries of patentable subject matter, neither the language
of the statute nor judicial decisions elaborating its meaning have explicitly
excluded "information" from patent protection. Arguably, such a limitation
is implicit in prior judicial decisions stating that the patent system protects
practical applications rather than fundamental new insights about the natural world.[13] The exclusion of information itself from patent protection is
also at least implicit in the statutory requirement that patent applicants
make full disclosures of information about their inventions, with no
restrictions upon public access to the disclosure once the patents issue.[14]

Patent claims on DNA sequences as "compositions of matter" give
patent owners exclusionary rights over tangible DNA molecules and constructs, but do not prevent anyone from perceiving, using, and analyzing
information about what the DNA sequence is. Once the patent issues, this
information becomes freely available to the world, subject only to the
inventor's right to exclude others from making and using the claimed compositions of matter. For patents on genes that encode therapeutic proteins,

the value of this exclusionary right over tangible compositions of matter has been sufficiently large (relative to the value of the information that spills over to the public through the patent disclosure) to motivate inventors to file patent applications rather than to keep the sequence secret.

By contrast, in the setting of high-throughput DNA sequencing, the informational value of knowing what the sequence is often exceeds the tangible value of exclusionary rights in DNA molecules and constructs, at least initially. This information base provides a valuable resource for future discovery, only part of which corresponds to those portions of the sequence encoding proteins. DNA molecules corresponding to the portions of the sequence that encode valuable proteins may ultimately prove valuable as tangible compositions of matter. But it might not be immediately apparent just which parts of the sequence encode which proteins, and what if anything makes those proteins valuable.

It is not obvious how an inventor might use patents to capture the value of the sequence under these circumstances. It may be difficult to draft claim language that covers the portions of the sequence that prove to have tangible value without claiming either too broadly (rendering the claim invalid because it covers similar sequences that have already been disclosed in the prior art) or too narrowly (rendering the claim easy to evade through minor changes in the molecule).[15] More important, claim language that is directed to tangible molecules fails to capture the informational value of knowing the sequence itself. If this informational value is large relative to the speculative value of tangible molecules corresponding to portions of the sequence, the more sensible strategy may be to sell access to a proprietary database of sequence information. So far, database subscriptions have been the principle source of revenue for most private firms involved in high-throughput DNA sequencing, although the same firms have also filed patent applications.

CLAIMING COMPUTER-READABLE INFORMATION

Another strategy that the PTO and the courts have yet to consider seeks to capture the informational value of DNA sequences through patent claims directed toward sequences stored in a computer-readable medium. An early example of this strategy is the patent application filed by Human Genome Sciences (HGS) on the sequence of the *Haemophilus influenzae* Rd genome.[16] This patent application was published eighteen months after its filing date under the terms of the Patent Cooperation Treaty, before it had issued anywhere in the world.[17] *Haemophilus influenzae* is a bacterial strain that causes ear and respiratory tract infections in humans. It was the first organism whose genome was fully sequenced, and the fate of the related patent applications may offer a preview of how the patent system will allocate patent rights in future genomic discoveries.[18] Human Genome Sciences filed a patent application setting forth the complete nucleotide sequence of the genome, identified as "SEQ ID NO:1."[19] The application concludes with a series of claims representing the invention to which HGS seeks exclusive rights. The first of these claims reads as follows:

> Computer-readable medium having recorded thereon the nucleotide sequence depicted in SEQ ID NO:1, a representative fragment thereof or a nucleotide sequence at least 99.9 percent identical to the nucleotide sequence depicted in SEQ ID NO:1.

It bears emphasizing that this is not the claim language of an issued patent. It is, in effect, the first item on the wish list of HGS for patent rights associated with the discovery of the *H. influenzae* genome.

This claiming strategy represents a fundamental departure from the previously sanctioned practice of claiming DNA sequences as tangible molecules. By claiming exclusionary rights in the sequence information itself, if stored in a computer-readable medium, HGS sought patent rights that would be infringed by information storage, retrieval, and analysis rather than simply by making, using, or selling tangible molecules.[20] Recently, the PTO issued a patent on this application with claims limited to tangible DNA molecules, but it remains to be seen what will happen when a persistent patent applicant pursues the matter on appeal to the Court of Appeals for the Federal Circuit.[21]

EXPANSIVE TREND OF CASE LAW

Recent decisions concerning the patentability of computer-implemented inventions may provide more guidance than prior decisions concerning the patentability of discoveries in the life sciences in predicting whether DNA sequence information stored in a computer-readable medium may be patented. The overall trend of decisions in the federal circuit is toward expansive interpretation of the scope of patent eligible subject matter— even for categories of inventions that prior decisions seemed to exclude from the protection of the patent statute—in order to make the patent system "responsive to the needs of the modern world."[22] The most conspicuous recent example of this trend was the 1998 decision in *State Street Bank & Trust* v. *Signature Financial Group* upholding the patentability of a computer-implemented accounting system for managing the flow of funds in partnerships of mutual funds that pool their assets.[23] This invention arguably fell within previously apparent judicial limitations that excluded mathematical algorithms and business methods from patent protection.[24] Yet the federal circuit minimized the first of these limitations, holding that it only excluded from patent protection "abstract ideas constituting disembodied concepts or truths that are not 'useful,'" and repudiated the second, insisting that "[t]he business method exception has never been invoked by this court, or [its predecessor], to deem an invention unpatentable," and that other courts that had appeared to apply the business method exception always had other grounds for arriving at the same decision.[25]

Rather than seeing the language of section 101 of the patent act, which permits patents to issue for "any new and useful process, machine, manufacture, or composition of matter," as a significant limitation on the types of advances that might qualify for patent protection, the federal circuit characterizes this language as a "seemingly limitless expanse," subject only to three "specifically identified . . . categories of unpatentable subject matter: 'laws of nature, natural phenomena, and abstract ideas.'"[26] In this environment, it is not obvious why DNA sequence information stored in a computer-readable medium would be categorically excluded from patent protection.

PTO GUIDELINES

Of course, DNA sequence information stored in a computer-readable medium is not the same thing as a computer-implemented business method, and it is certainly possible to define boundaries for the patent system that include the latter but not the former. Indeed, the PTO's "Examination Guidelines for Computer-Implemented Inventions" exclude data stored in a computer-readable medium from patent protection.[27] The guidelines distinguish between "functional descriptive material" (such as "data structures and computer programs which impart functionality when encoded on a computer-readable medium") and "nonfunctional descriptive material" (such as "music, literary works, and a compilation or mere arrangement of data [which] is not structurally and functionally interrelated to the medium but is merely carried by the medium").[28] Although functional descriptive material will generally fall within the statutory categories of patent-eligible subject matter, the guidelines state that nonfunctional descriptive material will generally not meet the statutory limitations:

> Merely claiming nonfunctional descriptive material stored in a computer-readable medium does not make it statutory. Such a result would exalt form over substance.[29]

DNA sequence information stored in a computer-readable medium seems to fall squarely within the PTO's definition of "nonfunctional descriptive material" that is "merely carried by" the computer-readable medium and is not functionally interrelated to it.

If the PTO continues to follow these six-year-old guidelines, it should reject claims to DNA sequence stored in a computer-readable medium. But if a disgruntled patent applicant appeals to the federal circuit, that court might well reverse the rejection. The distinction between tangible molecules and intangible information may do little work today in delineating the boundaries of patent eligibility in the face of recent decisions deemphasizing the importance of physical limitations in establishing the patentability of computer-implemented inventions. This shift in emphasis is particularly apparent in *AT&T* v. *Excel Communications*, in which the court explicitly declined to focus on the "physical limitations inquiry" that had played a central role in distinguishing between unpatentable mathematical algorithms and patentable computer- implemented inventions in its prior decisions.[30] Instead, the court asked "whether the mathematical algorithm is applied in a practical manner to produce a useful result."[31] This approach appears to merge the issue of patent eligibility with the issue of utility, opening the door to patent claims to information so long as it is "useful."

TRADITIONAL PATENT BARGAIN

If the federal circuit steps back from the momentum of its recent decisions expanding the boundaries of the patent system, it should not be persuaded by this argument. Patent claims to information stored in a computer-readable medium represent a fundamental departure from the traditional patent bargain. That bargain calls for free disclosure of information to the public at the outset of the patent term in exchange for exclusionary rights in particular tangible applications until the patent expires. Claims that are infringed by mere perception and analysis of the information set forth in the patent disclosure undermine the strong policy preventing patent applicants from restricting access to the disclosure once the patent has issued.[32] The limitation that the information be stored in a "computer-readable medium" offers scant protection for the public interest in free access to the informational content of patent disclosures. Scanning technologies arguably bring paper printouts of DNA sequence information within the scope of the claim language, an interpretation that would make copying the patent document itself an act of infringement. Even if the claim language is more narrowly interpreted to cover only electronic media, numerous Web sites post the full text of issued patents, including a Web site maintained by the PTO.[33] Any claim that would count these postings as acts of infringement simply proves too much.

That patents on information represent a departure from tradition may not be a sufficient ground to reject them in light of the increasing importance of information products to technological progress. Perhaps the traditional bargain of free disclosure of information in exchange for exclusionary rights in tangible applications doesn't make sense in this new environment. If the value of unprotectible information gained from high-throughput DNA sequencing is large relative to the value of tangible molecules that might be covered by established claiming strategies, patents that do not allow the inventor to capture the value of the information might not do enough to motivate investment in DNA sequencing. This may seem unlikely as an empirical matter, given the substantial investments that are being made in DNA sequencing in both the public and private sectors with no clear precedent for capturing the informational value of this investment through the patent system, but it is at least a logical possibility. A more plausible speculation is that inventors might forego the patent bargain if they are stuck with the traditional terms of that bargain, choosing instead to exploit their discoveries through restricted access to proprietary DNA sequence databases.

On the other hand, if the terms of the bargain are altered to allow patent holders to capture the informational value of their discoveries, the bargain becomes less attractive to the public. If the issuance of patents does not leave the public free to perceive and analyze the information disclosed in patent specifications, the public might be better off withholding patents and allowing others to derive the same information independently. Withholding patents makes particular sense if the efforts of the patent holder are not necessary to bring the information into the public domain. Much DNA sequence information is freely disclosed in the public domain, both by publicly funded researchers and by private firms. If a discovery is likely to be made and disclosed promptly even without patent incentives, there is little point in enduring the social costs of exclusionary rights.[34]

BRICKS-AND-MORTAR RULES FOR INFORMATION GOODS

There are sound policy reasons to be wary of permitting the use of the patent system to capture the value of information itself. The traditional patent bargain ensures that patenting enriches the information base, even as it slows down commercial imitation.[35] This balances the interests of the inventor in earning a return on past research investments against the interest of the larger public in avoiding impediments to future research. If patent claims could prevent the perception and analysis of information, this balance would tilt sharply in favor of patent owners.

More generally, patents are a form of intellectual property right that is particularly ill suited to the protection of information because there are so few safety valves built into the patent system that constrain the rights of patent holders in favor of competing interests of the public. Unlike copyright, patent law has no fair use defense that permits socially valuable uses to go forward without a license.[36] Contrary to the understanding of many scientists, patent law has only a "truly narrow" research exemption that offers no protection from infringement liability for research activities that are commercially threatening to the patent holder.[37] Nor is independent creation a defense to patent infringement. Unlike trade secret law, patent law has no defense for reverse engineering. The most important safety valve built into a patent, apart from its finite term, is the disclosure requirement that permits unlicensed use of information about the invention, as distinguished from the tangible invention itself.[38] But if patents issue that restrict the public from perceiving and analyzing the information, the claim effectively defeats that safety valve.

The foregoing discussion might seem to understate the implications for patents on DNA sequences of a principle that excludes information from patent protection. The DNA molecule itself may be thought of as a tangible storage medium for information about the structure of proteins. Cells read the information stored in DNA molecules to make the proteins that they need to survive in their environments, and they copy that information when they divide and reproduce. If DNA sequence information is not patentable when it is stored in a medium that is readable by computers, how can it nonetheless be patented when stored in a medium that is readable by living cells?

A quick answer is that information stored in a computer-readable medium is directed at the human observers who are the intended beneficiaries of the information spillovers that arise through patent disclosures. It is therefore *human*-readable information that must not be patented as such in order to maintain a balance between the exclusionary rights of patent holders and the rights of the public to use the disclosures that are the quid pro quo of those exclusionary rights. But humans can direct queries to DNA sequence information whether it is stored in molecular form or in electronic form. One might, for example, use DNA molecules as probes to detect the presence of a particular DNA sequence in a sample. This sort of molecular query has diagnostic and forensic applications as well as research applications. Researchers seeking to learn more about the functional significance of DNA sequence information are likely to query the information in both computer-readable and molecular form.[39] The distinction between computer-readable and molecular versions of DNA sequence is particularly difficult to maintain in the context of DNA array technology. DNA array technology involves immobilizing thousands of short oligonucleotide molecules on a substrate to detect the presence of particular sequences in a sample using specialized robotics and imaging equipment.[40] In effect, this technology enables people to use computers to perceive information stored in DNA molecules in a sample. When contemporary technology blurs the boundaries between computer-readable and molecular forms of DNA, what logic is there to drawing this distinction in determining the patent rights of DNA sequencers?

Perhaps the best argument for maintaining a distinction between DNA sequence information and DNA molecules at this point in the history of patents for genetic discoveries is consistency with tradition and precedent. Any categorical exclusion of DNA molecules from eligibility for patent protection would contradict the practice of the PTO and the courts for two decades and would undermine the precedent-based expectations of a patent-sensitive industry. On the other hand, allowance of patent claims to DNA sequence information stored in a computer-readable medium would extend patentable subject matter beyond what the PTO and the courts have recognized thus far, departing from a long-standing tradition of free access to the information disclosed in issued patents.

This analysis may seem stubbornly "bricks and mortar" in its focus on tangibility as the touchstone for protection, and therefore out of step with the needs of the modern information economy. If a significant portion of the value of DNA sequencing resides in the information that it yields, rather than in the molecules that correspond to that information, then perhaps we should not assume that investments in DNA sequencing will be forthcoming on the basis of an intellectual property system that limits exclusionary rights to tangible things and allows the value of the information itself to spill over to the general public.[41] At some point, we may need intellectual property rights that permit the creators of information products to capture the value of the information itself in order to motivate socially valuable investments. But if we have arrived at that point, then we need to look beyond the patent system for a suitable model. The patent system was designed to serve the needs of a bricks-and-mortar world, and it would be foolish to assume that it can meet the changing needs of the information economy simply by expanding the categories of subject matter that are eligible for patent protection.

NOTES

1. See Philip E. Ross, "The Making of a Gene Machine," *Forbes* (21 February 2000): 98; Paul Smaglik, "A Billion Base Pairs, Times Two," *Scientist* (6 December 1999); Francis S. Collins, "The Sequence of the Human Genome: Coming a Lot Sooner Than You Think," posted on the Internet at www.nhgri. nih.gov/ NEWS (accessed 10 January 2000); Justin Gillis and Rick Weiss, "Private Firm Aims to Beat Government to Gene Map," *Washington Post*, 12 May 1998, sec. 1, p. 1 (LEXIS News Library); Nicholas Wade, "Scientist's Plan: Map All DNA Within Three Years," *New York Times*, 10 May 1998, sec. 1, p. 1.

2. See Peter G. Gosselin, "Patent Office Now at Heart of Gene Debate," *Los Angeles Times*, 7 February 2000; Ralph T. King Jr., "Code Green: Gene Quest Will Bring Glory to Some; Incyte Will Stick With Cash," *Wall Street Journal* (10 February 2000), A1; Justin Gillis, "Md. Gene Researcher Draws Fire on Filings; Venter Defends Patent Requests," *Washington Post*, 26 October 1999, E1.

3. See Peter G. Gosselin, "Clinton Urges Public Access to Genetic Code," *Los Angeles Times* (February 2000).

4. See, for example, note, "Human Genes Without Functions: Biotechnology Tests the Patent Utility Standard," *Suffolk University of Law Review* 27 (1993): 1631; Philippe Ducor, "Recombinant Products and Nonobviousness: A Typology," *Computer and High Tech Law Journal* 13, no. 1 (1997); see Stanley Fields, "The Future Is Function," *Nature Genetics* 15, no. 325 (1997); Rebecca S. Eisenberg, "Structure and Function in Gene Patenting," *Nature Genetics* 15, no. 125 (1997); see Martin Enserink, "Patent Office May Raise the Bar on Gene Claims," *Science* 287, no. 1196 (2000); see Jon F. Merz et al., "Disease Gene Patenting Is a Bad Innovation," *Molecular Diagnosis* 2 (1997): 299–304 .

5. See Rebecca S. Eisenberg, "Patenting the Human Genome," *Emory Law Journal* 39, no. 721 (1990).

6. See, for example, *Amgen v. Chugai Pharmaceutical Co.*, 927 F2d 1200 (Fed Cir 1991); *Scripps Clinic & Research Found. v. Genentech*, 927 F2d 1565 (Fed Cir 1991); *Genentech v. The Wellcome Found.*, 29 F2d 1555 (Fed Cir 1994); *Hormone Research Found. v. Genentech*, 904 F2d 1558 (Fed Cir 1990); *Novo Nordisk v. Genentech*, 77 F3d 1364 (Fed Cir 1996); *Genentech v. Eli Lilly & Co.*, 998 F2d 931 (Fed Cir 1993); *Bio-Technology General v. Genentech*, 80 F3d 1553 (Fed Cir 1996); *Enzo Biochem v. Calgene*, (Fed Cir 1999).

7. Mark A. Lemley, "An Empirical Study of the Twenty-Year Patent Term," *American Intellectual Property Law Association Quarterly Journal* 22, no. 369 (1995).

8. See, for example, *Amgen* v. *Chugai Pharmaceutical Co.*, 927 F2d 1200 (Fed Cir), *cert. denied sub nom*; *Genetics Institute* v. *Amgen*, 502 US 856 (1991) ("A gene is a chemical compound, albeit a complex one. . . .")

9. U.S. Code Title 35 Sec. 101.

10. *Diamond* v. *Chakrabarty*, 447 US 303 (1980).

11. *Amgen, Inc.* v. *Chugai Pharmaceutical Co.*, 13 USPQ 2d (BNA) 1737 (D. Mass. 1990) ("The invention claimed in the '008 patent is not as plaintiff argues the DNA sequence encoding human EPO since that is a nonpatentable natural phenomenon 'free to all men and reserved exclusively to none. . . .' Rather, the invention as claimed in claim 2 of the patent is the 'purified and isolated' DNA sequence encoding erythropoietin.")

12. See, for example, *Merck & Co.* v. *Olin Mathieson Chemical Corp.*, 253 F2d 156 (4th Cir 1958) (upholding the patentability of purified Vitamin B-12).

13. See, for example, *Diamond* v. *Chakrabarty* ("Einstein could not patent his celebrated law that $E = mc^2$; nor could Newton have patented the law of gravity. Such discoveries are "manifestations of . . . nature, free to all men and reserved exclusively to none."); *Dickey-John Corp.* v. *International Tapetronics Corp.*, 710 F2d 329 (7th Cir 1983) ("Yet patent law has never been the domain of the abstract— one cannot patent the very discoveries which make the greatest contributions to human knowledge, such as Einstein's discovery of the photoelectric effect, nor has it ever been considered that the lure of commercial reward provided by a patent was needed to encourage such contributions. Patent law's domain has always been the application of the great discoveries of the human intellect to the mundane problems of everyday existence.")

14. U.S. Code Title 35, Sec. 112. 154(a)(4). See *In re Argoudelis*, 434 F2d 1390 (Ct Customs & Pat App 1970).

15. The language of patent claims defines the scope of the patent holder's exclusionary rights. U.S. Code Title 35, Sec. 112; *Ex parte Fressola*, 27 USPQ 2d (BNA) 1608 (Bd Pat App & Interf 1993).

A *broad claim* is a claim that has few limitations. One might, for example, seek a claim that covers any molecule that includes (or "comprises," in the vernacular of patent law) at least ten consecutive nucleotides from the disclosed sequence. If allowed, such a claim would be very broad in that it would be likely to cover any portion of the sequence that later proves to encode a valuable protein. But the breadth of the claim makes it more likely that it will be held invalid. The claim would be invalid if any previously disclosed DNA sequence included any ten consecutive nucleotides that were identical to any portion of the sequence disclosed in the patent appplication. The shorter the portion of the disclosed sequence that is necessary to establish infringement, the broader the claim. But the broader the claim, the easier it is to find "prior art" disclosures that would fall within the scope of the claim, rendering the claim invalid.

A *narrow claim* is a claim that has many limitations. One might, for example, claim the entire disclosed sequence as an isolated molecule. Since every element of the claim must be present in a competitor's product to establish infringement, a competitor who made a DNA molecule that included only a portion of the disclosed sequence corresponding to a particular protein would not be liable.

16. Nucleotide Sequence of the *Haemophilus influenzae* Rd Genome, Fragments Thereof, and Uses Thereof, WO 96/33276, PCT/US96/05320.

17. Patent Cooperation Treaty of June 19, 1970, Art. 21(2).

18. R. D. Fleischmann et al., "Whole-genome random sequencing and assembly of Haemophilus influenzae Rd." *Science* 269 (1995): 496–512.

19. The sequencing was done at The Institute for Genomic Research (TIGR), a private, nonprofit organization affiliated with Human Genome Sciences (HGS) at the time. Pursuant to an agreement between TIGR and HGS, patent rights in the *H. influenzae* genome were assigned to HGS.

20. The meaning of "computer-readable medium" could be quite broad. See *infra.*

21. U.S. Patent No. 6,355,450, issued March 12, 2002. Although the issued patent, in keeping with the original aspirations for claim language, is entitled "Computer-readable genomic sequence of Haemophilus influenzae Rd, fragments thereof, and uses thereof," the claim language ultimately allowed by the PTO is limited to tangible polynucleotide molecules and methods. An applicant whose claims have been rejected by a PTO examiner twice may appeal to the Board of Patent Appeals and Interferences, U.S. Code Title 35, Sec. 134, and an applicant who is dissatisfied with the decision of the Board of Patent Appeals and Interferences may appeal to the United States Court of Appeals for the Federal Circuit, Title 35, Sec. 141.

22. *AT&T Corp. v. Excel Communications, Inc.*, 172 F3d 1352 (1999).

23. 149 F3d 1368 (Fed Cir 1998), *cert. denied,* 119 S. Ct 851 (1999).

24. See, for example, *Gottschalk* v. *Benson,* 409 US 63 (1972); *Parker* v. *Flook,* 437 US 584 (1978); *Hotel Security Checking Co.* v. *Lorraine Co.* (1908).

25. 149 F3d at 455. Ibid. at 1462.

26. *AT&T Corp. v. Excel Communications, Inc.*, 172 F3d 1352 (Fed Cir 1999), citing *Diamond* v. *Diehr,* 450 US 175 (1981).

27. 61 Fed Reg 7478 (28 February 1996), posted on the Internet at <http://www.uspto.gov/web/offices/com/hearings/software/analysis/computer.html> (February 2000).

28. The focus on functional relationship between data and substrate echoes language from *In re Lowry*, 32 F3d 1579 (Fed Cir 1994), in which the Federal Circuit upheld the patentability of a data structure for storing, using, and managing data in a computer memory. In that case, the Board of Patent Appeals had reversed the examiner's rejection of the claims under U.S. Code Title 35, Sec. 101 as claiming nonstatutory subject matter, and the issue of patentable subject matter was therefore not properly before the court on appeal. Nonetheless, in its analysis of the remaining issues of patentability under Title 35, Secs. 102 and 103, the court drew a distinction between claiming information content and claiming a functional structure for managing information:

> "Contrary to the PTO's assertion, Lowry does not claim merely the information content of a memory. Lowry's data structures, while including data resident in a database, depend only functionally on information content. While the information content affects the exact sequence of bits stored in accordance with Lowry's data structures, the claims require specific electronic structural elements which impart a physical organization on the information stored in memory. Lowry's invention manages information. As Lowry notes, the data structures provide increased computing efficiency. Id. at 1583."

29. This qualification in the guidelines responds to a rhetorical question posed by Judge Archer in his dissenting opinion from the *en banc* decision of the Federal Circuit in *In re Alappat*, 33 F3d 1526 (1996). In that case a majority of the court upheld the patentability of a claim to a computer-implemented mechanism for improving the quality of a picture in an oscilloscope. Judge Archer cautioned against the potential implications of allowing patent claims on mathematical algorithms stored in computer-readable medium in his dissenting opinion, asking rhetorically whether a piece of music recorded on a compact disc or player piano roll would be patentable:

> Through the expedient of putting his music on known structure, can a composer now claim as his invention the structure of a compact disc or player piano roll containing the melody he discovered and obtain a patent therefor? The answer must be no. The composer admittedly has invented or discovered nothing but music. The discovery of music does not become patentable subject matter simply because there is an arbitrary claim to some structure.

33 F3d 1526, 1545, at 1554 (dissenting opinion).

30. 172 F3d 1352 (Fed Cir 1999)

31. Id. at 1352.

32. See *In re Argoudelis*, 434 F2d 1390, 1394-96 (CCPA 1970)(concurring opinion); *Feldman* v. *Aunstrup*, 517 F2d 1351, 1355 (CCPA 1975), *cert. denied*, 424 US 912 (1976).

33. See <http://www.uspto.gov/web/menu/pats.html>.

34. Normally the nonobviousness standard set forth at U.S. Code Title 35, Sec. 103 prevents the issuance of patents on inventions that are highly likely to be made independently by another inventor by excluding from patent protection inventions that would have been "obvious" to persons of ordinary skill in the field of the invention given the state of the art. This standard fails to serve this important function in the context of DNA sequencing because of decisions of the Federal Circuit upholding the patentability of newly identified DNA sequences discovered through routine work, so long as the prior art did not permit prediction of the structure of the DNA molecule. See *In re Bell*, 991 F2d 781 (Fed Cir 1993); *In re Deuel*, 51 F3d 1552 (Fed Cir 1995).

35. See 3 Chisum on Patents Sec. 7.01 ("Full disclosure of the invention and the manner of making and using it on issuance of the patent immediately increases the storehouse of public information available for further research and innovation and assures that the invention will be freely available to all once the statutory period of monopoly expires.")

36. U.S. Code Title 17 Sec. 107.

37. The quoted words are from the opinion of the Court of Appeals for the Federal Circuit in *Roche Prods., Inc.* v. *Bolar Pharmaceuticals, Inc.*, 733 F2d 858 (Fed Cir), *cert. denied*, 469 US 856 (1984). For a fuller discussion of the research exemption, see Rebecca S. Eisenberg, "Patents and the Progress of Science: Exclusive Rights and Experimental Use," *University of Chicago Law Review* 56 (1989): 1017.

38. The rule for determining the expiration date of a U.S. patent was changed in 1995 by the Uruguay Round Amendments Act, Pub. L. No. 103-465 (H.R. 5110). Prior to passage of that act, U.S. patents expired seventeen years after the date that they were issued, regardless of their application filing dates. The new rule, applicable to U.S. patents issued on the basis of patent applications filed after 8 June 1995, provides for expiration twenty years after their filing dates. U.S. Code Title 35, Sec. 154.

39. After sequencing DNA, researchers might analyze the sequence in computer-readable form to identify similarities to known sequences, and then analyze the sequence in cell-readable form to observe the functional significance of different portions of the sequence in a living cell or organism. They might, for example, use DNA molecules as probes to determine when and where an

organism expresses a particular portion of its DNA sequence, or they might induce a cell to express a particular DNA sequence in order to learn more about the protein that it encodes, or they might interrupt expression of a DNA sequence in an organism and observe the consequences in order to learn more about the functions of the corresponding protein. This sort of interaction between analysis of electronic information and observation of how cells use the information characterizes what in recent years has become known as "functional genomics" research. See Philip Hieter & Mark Boguski, "Functional Genomics: It's All How You Read It," *Science* 278 (1997): 601; Stanley Fields, "The Future Is Function," 15 *Nature Genetics* 15 (1997): 325.

40. See R. Ekins and F. W. Chu, "Microarrays: Their Origins and Applications," *Trends in Biotechnology* 17 (1999): 217; B. Sinclair, "Everything's Great When It Sits on a Chip: A Bright Future for DNA Arrays," *Scientist* 13 (24 May 1999): 18.

41. The classic argument for intellectual property is that exclusionary rights are necessary to motivate investments in the creation of goods that are costly to make initially, but cheap and easy to copy once someone else has made the initial investment. As growing volumes of information become freely available on the Internet, this argument seems to be overlooking significant incentives to create and disseminate information outside the intellectual property system. See, generally, Carl Shapiro and Hal Varian, *Information Rules: A Strategic Guide to the Network Economy* (Boston: Harvard Business School Press, 1999).

seven

DISCOVERIES, INVENTIONS, AND GENE PATENTS
DAVID B. RESNIK

Abstract: The discovery/invention distinction plays a key role in the debate about gene patents since only human inventions are treated as a legally patentable subject matter in the United States and other countries. Both proponents and opponents of gene patents seek objective support for their views by appealing to the difference between discovery and invention: proponents argue that specific genes or genetic technologies are human inventions (and hence patenable), while opponents argue that these items are products of nature (and hence not patentable). This essay argues that these appeals to the discovery/invention distinction are fundamentally misguided. The distinction between discovery and invention, like so many other key distinctions in bioethics and biomedical policy, is not purely objective (or value-free). Unreflective appeals to this distinction may beg the very question at hand or settle nothing at all. If we view the gene-patenting debate through this framework, we will be able to pay closer attention to the values that are at stake without being misled by queries about the "facts" relating to genetic discovery and invention.

INTRODUCTION

A key question in the controversy concerning human gene patenting is whether genes should be treated as a legally patentable subject matter. The United States legal system holds that only items that are the products of human ingenuity can be patented; naturally occurring items cannot be patented. Evidence for this legal doctrine can be found in the

U.S. Constitution as well as in several statutes and court decisions.[1] In a landmark case, *Diamond* v. *Chakrabarty* (1980), the U.S. Supreme Court ruled that life-forms can be patented if they are the products of human ingenuity. In this case the Court held that a hybridized bacterium that metabolizes oil is a human invention, even though the inventor, Chakrabarty, used a natural process (bacterial replication) to produce the modified organism. The Court recognized that discoveries of naturally occurring organisms are not patentable, but it held that Chakrabarty's bacteria are patentable because they resulted from human ingenuity and research.[2] In the wake of this ruling, the U.S. Patent and Trademark Office (PTO) has awarded patents on plants, animals, cells, organs, proteins, and genes. In assessing patent applications, the PTO has followed the *Chakrabarty* standard in deciding what counts as a patentable subject matter.[3]

In the ensuing debate over gene patenting, opponents of gene patents have argued that genes are a part of nature and therefore should not be patentable.[4] Proponents acknowledge that naturally occurring phenomena should not be patentable, but they argue that those genes that are human inventions should be patentable.[5] Since both sides of this debate agree that only inventions can be patented, much of the debate about gene patenting turns on the question of whether (or how) genes should be treated as human inventions.[6] An age-old philosophical distinction between invention and discovery therefore lies at the heart of this contemporary dispute concerning gene patenting.

This essay will critique the discovery/invention distinction in order to shed some light on this policy issue. I will argue that since every object, process, or phenomena that we label as "discovery" or "invention" results from some combination of human and natural activity, the distinction is based on a practical choice to focus on particular types or patterns of causation for particular purposes. Since this choice is based on norms and commitments relating to science, technology, and human ingenuity, how we make the discovery/invention distinction reflects particular interests and goals (or values). Where we decide to draw the line between discovery and invention depends more on our purposes in making the distinction than on objective demarcations or divisions. If we view the gene-patenting debate through this framework, we will be able to pay closer attention to the values that are at stake without being misled by queries about the "facts" relating to genetic discovery and invention.

THE DISCOVERY, INVENTION, AND THE GENE-PATENTING DEBATE

The discovery/invention distinction plays a crucial role in legal and policy debates about gene patenting in the United States and other countries. There are several ways that one could patent genetic technology under U.S. patent law.[7] Process patents allow the patentee to patent processes, methods, or techniques used in identifying, cloning, isolating, modifying, recombining, or sequencing genes. Composition of matter (COM) patents are patents on particular genes or combinations of genes, or DNA sequences, such as expressed sequences tags (ESTs) or single nucleotide polymorphisms (SNPs).

There are two different ways one might attempt to obtain COM patents on genes. COM patents could apply to the chemical structures of particular DNA sequences or to the functions of those structures in cell regulation, physiology, growth, and development.[8] So far, the PTO has treated genes as chemical structures and has granted patents on particular DNA sequences that have been isolated and purified.[9] For instance, a person who clones a gene could patent that cloned DNA sequence, while another person could patent a novel use of that sequence, for example, as a diagnostic marker for a particular genetic disease or condition. To date, most of the controversy surrounds COM patents as well as process patents that play a key role in genetic research and innovation.

The system of patent law in the United States (and many other countries) allows inventors to patent machines, processes, or compositions of matter (henceforth items) only if those items are:

1) Products of human ingenuity, that is, they are the result of an inventive step
2) Useful, that is, they have some specific practical or scientific use
3) Nonobvious, that is, they result of a nontrivial innovation
4) New or original, that is, they are not previously patented or described in a public document

To be patentable, an item must meet all of these conditions.[10] To obtain a patent, an inventor must submit an application that describes the item in enough detail that a person trained in the relevant technical field can reproduce that item. Patents do not allow inventors to have complete control over their inventions; they are nonrenewable rights granted to inventors to control the reproduction, sale, and use of their inventions for a specific length of time, for example, twenty years in the U.S.[11] Patent holders therefore do not own patented items in the way one might own a house or a car. Since patents do not give patent holders complete control over patented items, they are best viewed as a type of incomplete (or partial) commodification.[12]

In legal and policy debates about patents, it is usually assumed that one must determine whether an item meets the four conditions mentioned above in order to assess its patentability. Thus, the question "Is an item a product of human ingenuity?" is treated as conceptually prior to and distinct from the question "Is an item patentable?"[13] For the purposes of this essay, I will focus on only the first condition, human ingenuity, and I will not consider the other conditions.

The idea that the discovery/invention distinction should play a key role in patent law has evolved over time. For example, Article 1, Section 8, Clause 8 of the U.S. Constitution grants Congress the power to grant exclusive rights to authors and inventors in order to promote the progress of science and technology. But the Constitution does not clearly distinguish between rights in invention and rights in discovery. However, during the 1800s the courts and the Congress began to view that discovery/invention distinction as a key part of patent law and policy.[14] By the time the Supreme Court issued its opinion in *Chakrabarty*, the discovery/invention distinction was firmly in place.

The majority opinion in the *Chakrabarty* case reinforced the growing importance of the discovery/invention distinction in U.S. patent law. The Court held that "Laws of nature, physical phenomena, and abstract ideas have not been held patentable. . . . Thus a new mineral discovered in the earth or a new planet found in the wild is not patentable subject matter."[15] But the Court also held that "anything under the sun" which is "the result of human ingenuity and research" is patentable subject matter.[16] Chakrabarty's bacterium can be patented, according to the Court, because "his discovery is not nature's handiwork but his own invention."[17]

Proponents of gene patents argue that genes are human inventions and therefore patentable.[18] John Doll, the director of the PTO's Biotechnology Examination division, maintains that DNA sequences, though functionally identical to their natural counterparts, are products of human ingenuity and can be patented.[19] Although naturally occurring genes cannot be patented, inventors can patent DNA structures, DNA functions, or processes used in genetic technologies. Carl Feldbaum, the president of the Biotechnology Industry Organization, also supports this view.[20] Since most genes that are patented are functionally (though not structurally) identical to genes that occur in nature, a key point in defending the pro-patenting position is making sense of the claim that gene structures or functions are human inventions.[21]

Longtime critic of biotechnology Jeremy Rifkin has argued that genes are products of nature and should not be patented.[22] Rifkin's organization, the Foundation on Economic Trends, has played a key role in developing an organized opposition to gene patents.[23] Several years ago, Rifkin urged religious leaders to sign a petition against gene patents. On May 18, 1995, eighty representatives from various religious traditions and faith communities declared that "We believe that humans and animals are creations of God, not humans, and as such should not be patented as human inventions."[24] Although this statement mentions only humans and animals, the rest of the declaration implies that this prohibition on patenting should apply to all creations of God.[25]

The discovery/invention distinction also plays an important role in common heritage arguments against gene patents.[26] According to these arguments, genes are a type of public property (*res communis*), belonging to no one person. Genes are analogous to other resources common to all people, such as public lands, the oceans, or the atmosphere. In order to protect and preserve this common resource, specific forms of legal control, such as patenting, should not be allowed, according to this view.[27] Not all types of public property are products of nature. For example, bridges, works of art, museums, and roads are types of man-made public property. Thus, it is not necessary that one assume that genes are a part of nature in order to defend a common heritage argument against patenting. However, it is incumbent upon a defender of the common heritage argument to explain how genes constitute a type of *res communis*. The most plausible understanding of genes as *res communis* is to view genes as a type of natural resource, like plant and animal species, ecosystems, or the oceans.[28]

Critics of the common heritage argument assert that genes are not like natural resources because they cannot be equated with particular objects, processes, or locations. Since a genome is an abstraction instead of a con-crete thing, it is more like a map or schematic diagram than a piece of tissue, plant, or animal.[29] Although the planet Earth is a natural resource, a map of Earth is not. Hence, there is little basis for defending the idea that genes are a part of our common heritage, according to this view. Whether one agrees or disagrees with the common heritage argument against gene patenting, it is clear that the distinction between natural resources, which are discovered, and nonnatural resources, which are invented, plays a piv-otal role in understanding this issue as well.

To summarize this section, the distinction between discovery and invention plays a key role in debates about gene patents. In many instances, arguments for gene patenting are founded on the claim that genes are human inventions, while arguments against gene patenting assert that genes are human discoveries (or products of nature). Both sides assume that determining whether genes are discovered or invented will help us decide whether genes are patentable subject matter. In the next section, we will see whether this metaphysical distinction can do the conceptual work that has been assigned to it.

DISCOVERY VERSUS INVENTION

Most people have an intuitive grasp of the difference between "discovery" and "invention," and it is easy to produce paradigmatic examples of these concepts. Galileo discovered the moons around Jupiter, but Isaac Newton invented the reflecting telescope. Sardi Carnot discovered some principles of efficiency for engines, but James Watt invented an efficient version of the steam engine. Benjamin Franklin discovered that lightning is elec-tricity, but he also invented the lightning rod to protect buildings from lightning. William Roentgen discovered x rays, but Godfrey Hounsfield invented the Computerized Axial Tomographic (CAT) scanner. Albert Einstein discovered a relationship between matter, energy, and light ($E = mc^2$), but scientists working on the Manhattan Project took advantage of this relationship in inventing the atom bomb. A contractor working for George Bissell discovered oil in Pennsylvania, but Jesse Dubbs invented an improved process for refining crude oil into petroleum.[30]

If we move beyond these paradigmatic cases of discovery and invention, we see that they have some common features. First, we use the words "discover" or "invent" to describe human actions. However, descriptions of human conduct only make sense within a larger explanatory or interpretative framework.[31] When we say that someone "discovered" or "invented" something, we are offering an account of what this person did, which relates his action to a larger explanatory framework or context. For example, when we say that "Roentgen discovered x rays," we are relating his action to a larger framework that includes historical and sociological dimensions. The historical dimension of the framework includes previous work in the area as well as influential ideas, theories, concepts, and people. The sociological dimension includes the social, economic, and cultural conditions of his discovery. Since discovery and invention do not occur "in a vacuum," we must understand how and why particular actions occur in order to decide whether they count as "discoveries" or "inventions."[32]

Second, the explanatory framework we use involves a variety of human and natural causes, such as objects, structures, processes, forces, properties, and the like. Calling an item an "invention" or a "discovery" depends on how we conceptualize humanity's causal role in generating, designing, uncovering, describing, modifying, or reproducing the item.[33] When an item depends on human beings for its existence, we tend to view the item as an invention. Thus, the steam engine would probably not exist unless human beings designed and built it. When an item does not depend on human beings for its existence, when it has an independent existence, we tend to view the item as a discovery. For example, crude oil would have existed beneath the ground in Pennsylvania even if human beings had never drilled for oil in that location.[34] This rendering of the discovery/invention distinction helps us to understand why the courts often refer to discoveries as "products of nature" and refer to inventions as "products of human ingenuity."[35]

Since the explanatory frameworks we use in classifying items as "discoveries" or "inventions" may employ many different human and natural causes, it is important to be able to sort through the different causal patterns and forms. For our purposes, we can distinguish between three different types of causes used in explaining particular events:

A. Primary causes (or causal agents)
B. Contributing causes (or causal factors or conditions)
C. Counteracting causes (or countervailing causes)

On this scheme, all causes have statistical connections to their effects: primary and contributing causes tend to bring about specific effects; counteracting causes tend to prevent these effects.[36] For example, if lightning strikes a forest and starts a fire, the lightning would be a primary cause of the forest fire. Contributing causes would include environmental factors and conditions, such as dryness, wind speed and direction, and the density of vegetation. Counteracting causes could include rain, fire-fighting efforts, and natural barriers to the progression of forest fires, such as rivers and lakes.

There is often no objective difference between primary and contributing causes, especially when an event has many different causes. Whether we decide to call a particular cause a primary or contributing cause depends on our practical interests in explaining a particular event.[37] For example, in explaining why a person wrecked her car on an icy road, we may cite driver error, road conditions, road design, or poor tires as the primary cause of the wreck. A great deal depends on our interests in explaining the wreck. For instance, drivers' education instructors may focus on driver error, while tire salespeople may focus on poor tires, and highway engineers may focus on road design.

If we apply this analysis to discovery and invention, then it implies that human beings serve as primary causes in invention but that they only serve as secondary causes in discovery. For instance, the sentence "Franklin invented the lightning rod" can be interpreted as asserting that (1) Franklin was a primary cause of the existence of the lightning rod, even though (2) there were other secondary causes, such as previous discoveries and inventions, available materials, and so on. Although items that are discovered have an independent existence, human beings can also play a causal role in observing, purifying, reproducing, controlling, analyzing, modeling, or interpreting these items. Scientific discovery involves both human intervention and representation.[38] For instance, the sentence, "Roentgen discovered x rays" can be interpreted as asserting that (1) Nature is the primary cause of x rays, but (2) a human being (Roentgen) was a secondary cause in the production of x rays, since he observed, controlled, reproduced, and interpreted x rays. This analysis also allows us to make sense of codiscovery and coinvention: codiscovery or coinvention occurs when two or more people have equal causal roles in discovery or invention.

HARD CASES

The foregoing account of the discovery/invention distinction gives us a good grasp of paradigmatic (or "easy") cases of discovery and invention. However, if we move beyond the easy cases and look at harder ones, we find that it is often difficult to say whether something has been invented or discovered, and the difference between discovery and invention becomes unclear. It may also be difficult to decide who should be credited with inventing or discovering an item, even if we can settle questions about discovery and invention.[39] The history of science contains numerous examples of divisive priority disputes.[40]

Consider a case from the physical sciences. In order to make californium, scientists bombard lighter elements, such as plutonium, with hydrogen or helium nuclei. The resulting collisions can create small quantities of californium, which exist for no more than a few millionths of a second before they decay. In theory, violent explosions that occur in nature, such as supernovae, could create californium. Thus, it is possible for this substance to exist in nature, but it takes a great deal of human effort and ingenuity to reproduce quantities of the element sufficient for observation and experiment. So was californium discovered or invented? Are there two types of the substance, the naturally occurring version, which is discovered, and the version created in the laboratory?

Consider two examples from biological sciences. In 1928 British bacteriologist Alexander Fleming recognized that molds contain a substance that kills bacteria. In 1940 two British scientists, Howard Florey and Ernst Chain, isolated and purified the substance, which became known as penicillin.[41] Although penicillin can exist in nature, human ingenuity is required to transform it into a medically useful compound. So is penicillin an invention or a discovery? Could the purified version of penicillin be an invention while the naturally occurring version is a product of nature?

One can tell a similar story about erythropoietin, an enzyme that stimulates the production of red blood cells. Over a decade ago, the Genetics Institute Inc. isolated and purified a small amount of erythropoietin from human urine. Did the Genetics Institute invent this protein? The U.S. PTO thought so: in 1987 it granted a patent on erythropoietin to the company.[42] However, one might also argue that even the purified form of this compound could exist in nature without being made by human beings.

These three examples, one from physics and two from biology, raise the question of whether we can treat two versions of the same thing as different things. If a product occurs in nature but also can be created artificially, does it make sense to say that one product is invented while the other is discovered? If so, what amount or type of human ingenuity is required to transform a naturally occurring thing into an invention? In the three examples mentioned above, human beings artificially reproduced naturally occurring substances. According to the PTO, an artificially reproduced item, such as a purified chemical, is a human invention.[43] Thus, one might be tempted to argue that artificially reproduced items are human inventions. But the matter is not as simple as this, since we tend to think that some artificially reproduced items, such as salt, sugar, oxygen, water, electricity, hydrogen, alcohol, combustion, and children are not human inventions. But is there an objective difference between artificial oxygen and an artificial antibiotic?

Before concluding this section, I would like to consider two more difficult cases. In the 1980s Harvard scientist Philip Leder transferred oncogenes into the germ line of laboratory mice. The progeny of these mice contain several genes that give them an increased risk of developing cancer, and they are therefore useful animal models for studying the development and prevention of cancer. These mice already existed in nature, but human ingenuity was required to develop the Harvard OncoMouse. In 1988 the PTO gave a patent to Leder for his transgenic mouse.[44] Did he invent the whole mouse or only a part of it? According to the PTO, the patent applied to the whole mouse, not just to the mouse's modified genes or tissues. Once again we are faced with the basic question: What type of human intervention is required to transform a naturally occurring item into an invention? We frequently modify the parts of naturally occurring items, such as wood, rocks, oil, fields, rivers, and mountains, without claiming to invent the whole item. I think must people would agree that a person who carves a flute out of a stick of wood invents part of the item but not the whole item. One part of the item—its design—is a human invention, but another part—its material—is not. If the whole flute is not an invention, then does it make sense to say that the whole mouse is an invention? Is there an objective difference between an artificial mouse and a wood flute?

Finally, consider the discovery of the structure of DNA. Historians have credited James Watson and Francis Crick with this important discovery, but their work did not occur in a "vacuum." Many other scientists made important contributions to this discovery. For example, Rosalind Franklin provided important evidence that helped to confirm the model, and many other scientists, such as Phoebus Levine, Oswald Avery, Andre Boivin, Erwin Chargoff, and Linus Pauling, provided ideas and information that were used in developing the model.[45] One might argue that many people helped to discover the structure of DNA, not just the two people who won the Nobel Prize for their work. Sometimes we do give credit to many people in science, not just to one or two key players. For example, many scientific papers today list six or more authors. Some have listed as many as several hundred authors. The same paper could have two or six authors, depending on how one decides to allocate credit. Likewise, the same item could be discovered by one or many people, depending on how one apportions responsibility. The assignment of credit in science is not a purely objective matter but depends on one's values, goals, and purposes in making the assignment.[46]

A PRAGMATIC APPROACH TO DISCOVERY AND INVENTION

In all of these examples we have just considered, there are many different human and natural causes, and it is difficult to say who or what is the primary cause of the item in question. These cases all involve some degree of human effort and ingenuity in transforming, designing, recreating, purifying, controlling, observing, or modifying a product of nature. If we think of discovery and invention as marking opposite ends of a line (see figure 1), then the hard cases are somewhere in the middle of this continuum.

Figure 1: Discovery versus Invention

Discovery	Hard Cases	Invention

Field mouse	Transgenic mouse	Mickey Mouse
Gold	Californium	Gold jewelry

Paradigmatic cases of assigning credit in science include obvious examples of single authorship or clear cases where a person is not an author. Hard cases are inventions or discoveries with multiple authors, such as the discovery of DNA or the invention of the television.

In all of these hard cases, we must mark some boundaries between invention and discovery or authors and nonauthors. To draw these lines, we must go beyond the objective facts of the situation and consider the values that are at stake. Most of the controversial cases, such as gene patenting, are hard cases. Moreover, we cannot appeal to the "objective" facts of causation in these hard cases, since the decision to view something as a primary cause is often a practical choice we make in order to promote specific values (interests, goals, or purposes). To settle these hard cases, we, therefore, need a careful analysis of the values at stake in using the terms "discovery" or "invention" in various contexts, since an appeal to objective facts can take us only so far. Another way of putting this point is to say that these distinctions depend on pragmatic features of language.

The distinction between "discovery" and "invention" is thus more like the distinction between "human" and "nonhuman" than the distinction between "positive charge" and "negative charge." Whether a particular object has a positive or negative charge depends on specific properties that have an independent existence. Electrons would have a negative charge even if human beings never discovered this fact. Whether a particular being is a human depends on who is making this claim and why. According to some moral theories, all members of the species *Homo sapiens* are human beings; according to other theories, only sentient members are fully human.[47]

The distinction between discovery and invention does not mark a purely factual or descriptive boundary. In philosophical parlance, the distinction has a normative basis: whether we call something an "invention" or "discovery" often says more about our values than the thing itself. In order to understand or interpret the distinction, one must therefore know who is making the distinction and why they are making it. It is plausible that two individuals (or societies) with different values could agree on the facts surrounding a particular item yet disagree about whether it should be called an invention or a discovery or who invented or discovered it.

If the distinction between invention and discovery does not mark an objective or factual boundary, then it may be counterproductive and even misleading to assume that it is an objective distinction. Instead of focusing on complex metaphysical or causal issues that may have no definite answer, we would do better to thoughtfully address the different values at stake. The following is a brief list of some of the values at stake in debates about invention and discovery:

- Scientific/technological. One might be concerned about the distinction in order to promote scientific or technological advancement.
- Scholarly. One might be concerned about the distinction in order to understand and interpret the history, philosophy, psychology, and sociology of science and technology. Was Leonardo Da Vinci a scientist, inventor, or both? Did Benjamin Franklin discover electricity? Did Galileo invent the telescope?
- Environmental/ecological. One might be concerned about the distinction in order to understand the role of human activity in the ecosystem. How has technology affected the climate?
- Religious/theological. One might be concerned about the discovery/invention distinction in order to understand humankind's proper relation to God's creation. Are there aspects of God's creation that human beings should not change or destroy? Are there moral or theological limits to human creativity and ingenuity? Do some activities represent hubris or arrogance?
- Moral/political. One might be concerned about the distinction in order to understand the nature of science and technology and their relation to human dignity and other values. How does calling something an "invention" affect its moral status? If we could design a human being, should we call him an invention?
- Economic. One might be concerned about this distinction in order to understand the role of human activity in economic development. How do scientific and technical innovations affect productivity and growth? What is the best way to encourage investment in science and technology?
- Law and public policy. Finally, one might be concerned about this distinction in order to address legal questions pertaining to the interpretation of statutes, the significance of court cases, or the moral justification of various regulations or policies. Should the PTO consider gene fragments to be discoveries or inventions?

It almost goes without saying that these different values may sometimes conflict, and the recent history of biotechnology illustrates this conflict. In the debate about patenting biological materials, such as animals, plants, and cell lines, we have seen conflicts of scientific, religious, economic, environmental, and moral/political values.[48] It is ironic that both sides in these debates have argued that the objective facts about invention and discovery support their case. For instance, both sides in the gene-patenting debate agree that products of nature are not patentable, but they disagree on the question of whether genes are products of nature or human inventions.[49] If my approach to this dispute is correct, then both sides (and some courts) are deeply misguided: the gene-patenting debate cannot be settled by appealing to some "objective" distinction between invention and discovery; it can only be resolved by carefully assessing and weighing the different values at stake. The dispute therefore resembles some other ethical/legal disputes in medicine, such as abortion. The distinction between discovery and invention, like the distinction between human and nonhuman, is not a purely objective distinction. In calling something a discovery (or an invention), we are making a judgment that has important implications for human values.

GENETIC INVENTIONS AS HARD CASES

Let's take what points I have made in the previous two sections and apply them to genetic technology. In the previous two sections, I have argued that although we can easily agree on paradigmatic cases of invention and discovery, we have difficulty distinguishing between hard cases. This is because the distinction between discovery and invention is not purely objective. To resolve these hard cases, we must carefully weigh and assess the values at stake. In this section I would like to apply these ideas to genetic technology.

To apply these ideas, we need to identify paradigmatic cases of discovery and invention in genetics, keeping in mind that there are different ways one might attempt to patent genes. Once we have these cases identified, we can address some hard cases. However, since genetics is a new and rapidly developing science, I suspect that it may be difficult to find paradigmatic cases of discovery and invention, since people may not agree about them. Be that as it may, we should at least attempt to establish benchmarks for discovery and invention.

Let me begin by proposing some paradigmatic cases of discovery in genetics. These include:

- Research that identifies the specific biochemical structures of naturally occurring genes, such as sequencing the approximately one hundred mutations that cause cystic fibrosis (CF).
- Research on some natural functions of genes, such as research on protein synthesis, programmed cell death, epigenesis, and cell regulation.[50]
- Research on naturally occurring processes in genetics, such as replication, transcription, translation, error correction, or recombination.

I will also propose some paradigmatic cases of invention in genetics. These include:

- Research that modifies substantially the biochemical structures of naturally occurring genes, gene fragments, or genomes for uses in biotechnology or genetic engineering, such as making expressed sequences tags (ESTs), attaching genes to viral vectors, or inserting genes in naturally occurring genomes. (For objections, see below.)
- Developing diagnostics tests for genetic diseases, which assign genes a functional role in disease prediction, prevention, and treatment. For example, BRCA 1 and 2 mutations can function as predictors of breast and ovarian cancer. (For objections to this example, see below.)
- Developing a process for sequencing, mapping, or cloning genes, such as gel electrophoresis or the polymerase chain reaction (PCR).

However, many types of genetic research fall in the gray area between invention and discovery. Some of these include:

- Gene structures or gene fragments that have been artificially reproduced or cloned.
- Genes functions that have not been substantially modified.
- Artificially reproduced natural processes, such as DNA replication under laboratory conditions (i.e. cloning) or genetic recombination in the laboratory.

In addition, one might challenge some of my proposed paradigmatic cases of invention. For instance, one might argue that a gene that is modified by attaching it to a vector has not been substantially modified. One might also argue that an EST is still a natural product that has not been modified enough to make it an invention. Finally, one might argue that using a gene in diagnosis is not an invention but a kind of discovery, that is, that a particular gene is associated with a particular disease.[51]

These considerations imply that genetics presents us with many hard cases for the discovery/invention distinction (see figure 2). In genetics, there are many kinds of cases where the objective "facts" do not determine whether we should classify items as products of nature or as products of human ingenuity. To resolve these difficult cases in genetics, we must weigh and assess the different values at stake.

Figure 2: Genetic Discovery vs. Genetic Invention

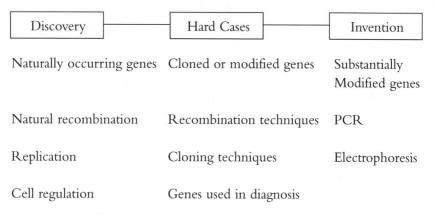

Discovery	Hard Cases	Invention
Naturally occurring genes	Cloned or modified genes	Substantially Modified genes
Natural recombination	Recombination techniques	PCR
Replication	Cloning techniques	Electrophoresis
Cell regulation	Genes used in diagnosis	

To explore these hard cases further, let's consider one hard case listed above, cloned genes. Proponents of gene patents, especially those associated with the biotechnology industry, have argued that cloned genes are human inventions.[52] According to this view, cloning a gene is not at all like making a photocopy of a text. To clone genes, scientists use a variety of biochemical techniques and compounds to isolate, purify, and copy DNA. In order to copy DNA, scientists use a DNA probe to produce a complementary strand of DNA (cDNA). By the time this process is complete, the end result—the cloned gene—is not the same as the gene that occurs in nature. It has been copied many times and is actually a form of cDNA, rather than the naturally occurring DNA. The cDNA may also be attached to other strands of DNA or viral vectors.[53] Scientists can use the same techniques to produce DNA fragments, such as SNPs or ESTs, or disease-causing genes. Cloned genes are like new translations or versions of a text rather than photocopies, one might argue.

On the other hand, opponents of gene patents have argued that cloned genes are still products of nature, not human inventions. Even though it takes a great deal of effort to find genes and remove them from nature, cloned genes are still discoveries, not inventions. A cloned gene is no more a human invention than sugar is a human invention. In both cases, one might argue, the same item occurs in nature and under artificial conditions. People do not design or create sugar; they only reproduce it through artificial means. To be a genuine invention, something must be sufficiently modified so that it is no longer the same as the thing that occurs in nature. Candy is a human invention, but sugar is not.

The counterreply to this objection is that a cloned gene has been sufficiently modified to be different from a naturally occurring gene. A cloned gene is more like refined sugar, which one might argue is a human invention. The rejoinder to this counterreply is that a cloned gene is more like maple syrup than refined sugar; it has not been modified enough to make it a product of human ingenuity.

If we continue following the debate on this path, we will sink further in murky metaphysical waters concerning identity and individuation. When are two items the same thing? When are they two distinct things? Philosophers have been debating these questions for centuries, and they have not developed a widely accepted and uncontroversial theory of identity.[54] So long as one does not hold that there is one "objective" metaphysical theory, then our answers concerning the identity and individuation of genes will reflect our values, goals, and purposes as well.[55] Our metaphysical assumptions depend on norms and commitments, not the "facts."[56]

These difficulties in finding a way to settle hard cases like cloned genes all add more evidence to support my claim that there are hard cases for the discovery/invention distinction in genetics that cannot be settled by appealing to the objective "facts." To determine whether a genetic structure, function, or process is (or is not) a human invention, we must weigh and assess the values, norms, and purposes that have some bearing on the issue. This type of assessment is inherently goal oriented and pragmatic.[57]

CONCLUSION: HOPE FOR A
MEANINGFUL DEBATE

Let's summarize the argument to this point. The discovery/invention distinction plays a key role in the debate about gene patents since only human inventions are treated as a patentable subject matter according to U.S. patent law. Both sides in the patenting debate have appealed to this distinction as if it would lend objective support to their position: defenders of gene patents argue that specific genes or genetic technologies are human inventions, while opponents argue that these items are products of nature. According to the view I have defended here, this appeal to the discovery/invention distinction as an objective determinant of patentability is fundamentally misguided. The distinction between discovery and invention, like so many other key distinctions in bioethics and biomedical policy, is not purely objective (or value-free).[58] Indeed, unreflective appeals to this distinction may beg the very question at hand or settle nothing at all.

An analogy with the abortion debate may be useful here. The distinction between "genetic inventions" and "genetic discoveries" is in many ways like the distinction between "human" and "nonhuman." Just as it makes little sense to expect that an objective distinction between human and nonhuman can settle the abortion issue, it makes little sense to expect that appeals to an "objective" distinction between invention and discovery can settle the gene-patenting debate. Defining what it means to invent a gene (or genetic technology) is one of the key issues in the genetic-patenting debate; likewise, defining what it means to be human is one of the key issues in the abortion debate.[59]

In order to settle cases straddling the line between invention and discovery (or human and nonhuman), we must weigh and assess the values, interests, and goals that are at stake. Ergo, we cannot appeal to the "facts" of human invention and genetic technology in order to resolve debates about the patentability of genes and genetic technologies, since the "facts" pertaining to discovery and invention are the focal point of the controversy. To resolve these debates, we must assess and weigh the different interests, goals, and values that play a role in forming our judgements about invention and discovery in genetics.

To a certain extent, this process of examining and weighing different values is already occurring in our public and scholarly debates about gene patents. Leaders of the biotechnology industry have defended patents on economic grounds, while many religious leaders have opposed them on theological and moral grounds. Scientists, for the most part, remain divided on the question of gene patenting.[60] Many individuals and organizations have publicly expressed conflicting views on gene patenting.[61] Moreover, one might argue that our entire intellectual property system is designed to strike some kind of balance between different interests and values.[62]

Thus far, the courts and the PTO have not come to terms with all of these different values in making decisions about the patentability of organic materials, including genes. This is not to say, however, that these institutions have not considered any values. On the contrary, the legal system has assessed intellectual property rights with an eye toward economic values. In deciding whether to award patents, the legal system has considered the impact of awarding patent rights on free trade, industrial innovation, and economic development.[63] Although the courts often discuss impacts on scientific and technical innovation, these impacts are viewed as relevant only insofar as they affect free trade and the economy; scientific and technical innovation is not treated as inherently valuable.[64] In the *Chakrabarty* decision, the Supreme Court held that it was not qualified to base its ruling on noneconomic values, such as moral, environmental, or religious values, since these related to matters of high policy that should be decided by Congress.[65]

According to the view defended here, the decision to not consider these other values is fundamentally misguided. By focusing on economic values, the legal system does not therefore render decisions that are objective or value-free. On the contrary, it produces outcomes based on privileging some values and marginalizing others. If we recognize that the debate about gene patenting involves fundamental conflicts of values, then we should find a way of promoting genuine dialogue and discussion without excluding different groups that have a stake in this issue.[66] We need to find a way of giving a fair hearing to the religious, moral, environmental, and other values that have some bearing on the gene-patenting debate.[67]

Even if one agrees that these other values should be given a fair hearing, one still might argue that the courts and the PTO are not the proper venues for discussing noneconomic values in gene patenting. These institutions, one might argue, are only qualified to address narrow questions of economic value relating to commerce, free trade, innovation, and product development. Perhaps this is the case. This is a point I will not debate here.[68]

However, there are larger issues at stake here than the scope of the PTO and the courts. Since gene patenting is a controversial public policy issue, we should find a way of giving serious consideration to different sides of this debate through a democratic process. Since the courts and the PTO have not given due consideration to noneconomic values, the legislative or executive branches of government should address and discuss these values through public forums. We need these public debates in order to resolve some of the hard questions about genetic discovery and invention that the legal system has not tackled and to decide which genes and genetic technologies can be patented and why. Without this public deliberation, only one point of view, the economic one, will be considered.

NOTES

1. Frank H. Foster and Robert L. Shook, *Patents, Copyrights and Trademarks* (New York: John Wiley and Sons, 1989).

2. *Diamond v. Chakrabarty*, 447 US 303 (1980).

3. John J. Doll, "The Patenting of DNA," *Science* 280 (May 1998): 689–93; Rebecca Eisenberg, "Patenting Organisms," in *Encyclopedia of Bioethics*, rev. ed. (New York: Simon and Schuster, 1995): 1911–14.

4. Jeremy Rifkin, *The Biotech Century: Harnessing the Gene and Remaking the World* (New York: Penguin, 1998); Jon Merz et al. "Disease Gene Patenting Is Bad Innovation," *Molecular Diagnosis* 2, no. 4 (1997): 299–304; Jon Merz and Mildred Cho, "Disease Genes are Not Patentable: A Rebuttal of McGee," *Cambridge Quarterly of Healthcare Ethics* 7 (1998): 425–28; David Saperstein, "Press Release on Gene Patenting," Religious Action Center of Reformed Judaism (18 May 1995): 1; Richard Land and Ben Mitchell, "Patenting Life: No," *First Things* 63 (May 1996): 20–22; Richard Stone, "Religious Leaders Oppose Patenting Genes and Animals," *Science* 268 (1995): 5214.

5. G. Poste, "The Case for Genomic Patenting," *Nature* 378 (1995): 535; L. Guenin, "Norms for Patents Concerning Human and Other Life Forms," *Theoretical Medicine* 17 (1996): 279–314; David Resnik, "The Morality of Human Gene Patents," *Kennedy Institute of Ethics Journal* 7, no. 1 (1997): 43–61; Glenn McGee, "Gene Patents Can Be Ethical," *Cambridge Quarterly of Healthcare Ethics* 7 (1998): 417–21; J. Tribble, "Gene Patents—A Pharmaceutical Perspective," *Cambridge Quarterly of Healthcare*

Ethics 7 (1998): 429–32; John Doll, "The Patenting of DNA," *Science* 280 (1998): 689–93; Gillian Woollett and O. Hammond, "An Industry Perspective on Gene Patenting, in *Perspectives on Gene Patenting*, ed. Audrey Chapman (Washington, D.C.: American Association for the Advancement of Science, 1999), 43–50.

6. Glenn McGee, "Gene Patents Can Be Ethical," *Cambridge Quarterly of Healthcare Ethics* 7 (1998): 417–21; Mark Sagoff, "DNA Patents: Making Ends Meet," in *Perspectives on Gene Patenting: Religion, Science, and Industry in Dialogue*, ed. Audrey Chapman (Washington, D.C.: American Association for the Advancement of Science, 1999), 245–68; Ted Peters, *Playing God?: Genetic Determinism & Human Freedom* (New York: Routledge, 1997).

7. David Resnik, "The Morality of Human Gene Patents"; Rebecca Eisenberg, "Structure and Function in Gene Patenting," *Nature Genetics* 15, no. 2 (1997): 125–30.

8. By allowing these two types of patents on genes—structure patents and function patents—the legal system appears to take a stand on the controversial issue of reductionism in the philosophy of biology. According to reductionists, genes are identical to DNA sequences; according to the holists, genes cannot be equated with chemical structures. Most holists assert that genes cannot be reduced to DNA because they have functional properties that cannot be entirely explained by their chemical composition. To explain these functional properties, one must understand the gene's role in larger organic processes, such as inheritance, development, growth, cell regulation, physiology, and behavior. The legal system appears to embrace reductionism as well as holism by allowing patents on both structures and functions. This is a complex question that I will not discuss further in this essay. For further discussion of this particular issue, see Eisenberg, "Structure and Function in Gene Patenting." For further discussion of reductionism in biology, see A. Rosenberg, *Instrumental Biology or the Disunity of Science* (Chicago: University of Chicago Press, 1994); Kenneth Schaffner, *Discovery and Explanation in Biology and Medicine* (Chicago: University of Chicago Press, 1993).

9. Doll, "The Patenting of DNA."

10. Foster and Shook, *Patents, Copyrights, and Trademarks*; Doll, "The Patenting of DNA."

11. Foster and Shook, *Patents, Copyrights, and Trademarks*.

12. Margaret Jane Radin, *Contested Commodities* (Cambridge, Mass.: Harvard University Press, 1996).

13. Sagoff, "DNA Patents: Making Ends Meet."

14. Foster and Shook, *Patents, Copyrights and Trademarks*.

15. *Diamond* v. *Chakrabarty*, 447 US 308-309 (1980).

16. Ibid., 310.

17. Ibid.

18. McGee, "Gene Patents Can Be Ethical."

19. Doll, "The Patenting of DNA."

20. Reginald Rhein, "Gene Patenting Crusade Moving from Church to Court," *Biotechnology Newswatch* (5 June 1995): 21.

21. Sagoff, "DNA Patents: Making Ends Meet"; McGee, "Gene Patents Can Be Ethical."

22. Rifkin, *The Biotech Century*.

23. Edmund Andrews, "Religious Leaders Prepare to Fight Patents on Genes." *New York Times*, 13 May 1995, N1, L1.

24. "Joint Appeal Against Human and Animal Patenting," quoted in C. Mitchell, "A Southern Baptist looks at Patenting Life," in *Perspectives on Gene Patenting*, ed. Audrey Chapman (Washington, D.C.: American Association for the Advancement of Science, 1999), 169.

25. Peters, *Playing God*; R. Cole-Turner, "Genes, Religion and Society: The Developing Views of the Churches," *Science and Engineering Ethics* 3, no. 3 (1997): 273–88.

26. Arthur Caplan, "What's So Special about the Human Genome?" *Cambridge Quarterly of Healthcare Ethics* 7 (1998): 422–24.

27. Pilar Ossori, "Common Heritage Arguments against Patenting Human DNA," in *Perspectives on Gene Patenting*, ed. Audrey Chapman (Washington, D.C.: American Association for the Advancement of Science, 1999): 89–108.

28. Ibid.

29. Eric Juengst, "Should We Treat the Human Germ-Line as a Global Human Resource?" in *Germ-line Intervention and Our Responsibilities to Future Generations*, ed. Emmanuel Agius and Salvino Busuttil (London: Kluwer Academic Press, 1998): 85–102.

30. Trevor Williams, *The History of Invention* (New York: Facts on File Publications, 1987).

31. Alexander Rosenberg, *Philosophy of Social Science*, 2d ed. (Boulder, Colo.: Westview Press, 1995).

32. Thomas Kuhn, *The Structure of Scientific Revolutions*, 2d ed. (Chicago: University of Chicago Press, 1970); Aharon Kantorovich, *Scientific Discovery* (Albany, N.Y.: State University of New York Press, 1993).

33. Kantorovich, *Scientific Discovery*.

34. Ibid.

35. It is important to note that this account of the discovery/invention distinction assumes that it is possible for some items to exist independent of human activity. If nothing existed independent of human activity, that is, if all objects, processes, properties, and phenomena were human creations or constructions, then

the distinction between "invention" and "discovery" would be moot. In a profound sense, everything would be a human invention (or construction) and nothing would be natural. In order to interpret the discovery/invention distinction, it is therefore necessary to assume some form of scientific realism. Thus, if one claims that Galileo discovered Jupiter, then one is committed to believing that Jupiter exists independent of Galileo. At a bare minimum, scientific realism involves the claim the objects exist independent from human theories, concepts, perceptions, values, and beliefs. One need not assume an all-encompassing (or global) version of scientific realism, however, in order to claim that a particular thing is a discovery. Arguments for scientific realism might only make sense within a localized context. For example, one might argue that electrons are real but that strings are not. Likewise, one might argue that some things can be discovered while other things cannot be discovered; they can only be invented. For more on scientific realism, see R. Boyd, "The Current Status of Scientific Realism, in *Scientific Realism*, ed. Jarrett Leplin (Berkeley: University of California Press, 1984): 41–82. For further discussion of discovery, see also Scott Kleiner, *The Logic of Discovery* (Dordrecht: Kluwer Academic Publishers, 1993).

36. Paul Humphreys, *The Chances of Explanation* (Princeton: Princeton University Press, 1989).

37. Baruch Brody, "Towards an Aristotelian Theory of Scientific Explanation," *Philosophy of Science* 39 (1972): 20–31; Wesley Salmon, *Four Decades of Scientific Explanation* (Minneapolis: University of Minnesota Press, 1989).

38. Ian Hacking, *Representing and Intervening* (Cambridge: Cambridge University Press, 1983).

39. David Magnus, "Ethical Implications of Gene Patenting," Association for Bioethics and the Humanities meetings, Philadelphia, Pennsylvania, 29 October 1999.

40. Robert Merton, *The Sociology of Science* (Chicago: University of Chicago Press, 1972); Kuhn, *The Structure of Scientific Revolutions*.

41. Williams, *The History of Invention*.

42. Sagoff, "DNA Patents: Making Ends Meet."

43. Doll, "The Patenting of DNA."

44. Peters, *Playing God*.

45. Ernst Mayr, *The Growth of Biological Thought* (Cambridge, Mass.: Harvard University Press, 1982).

46. David Resnik, *The Ethics of Science* (London: Routledge, 1998).

47. Bonnie Steinbock, *Life Before Birth* (Oxford: Oxford University Press, 1992).

48. Richard Gold, *Body Parts* (Washington, D.C., 1996); Radin, *Contested Commodities*; John Evans, "The Uneven Playing Field on the Dialogue on Patenting," in *Perspectives on Gene Patenting*, ed. Audrey Chapman (Washington, D.C.: American Association for the Advancement of Science, 1999).

49. Chapman, *Perspectives on Gene Patenting*; Evans, "The Uneven Playing Field on the Dialogue on Patenting."

50. I am assuming that we can distinguish between natural and artificial functions. For the purposes of this essay, artificial functions are functions that result from human ingenuity; natural functions are functions that exist independently. For example, a natural function of the heart is to pump blood; it would have this function even if human beings never existed. An artificial function of the heart is to produced sounds that can be used by a physician to diagnose heart disease. The heart would not have this function if people never existed or never learned to use heart sounds in diagnosis.

51. Merz and Cho, "Disease Genes Are Not Patentable: A Rebuttal of McGee."

52. Woollett and Hammond, "An Industry Perspective on Gene Patenting."

53. Doll, "The Patenting of DNA."

54. For a review of different theories of identity, see Hamlyn (1984).

55. I am assuming that, for the purposes of this discussion, that our metaphysical theories are not objective. The decision to adopt a particular metaphysical view (or framework) depends on one's goals and values. For more on this view, see Quine 1969, Putnam 1981.

56. It is worth noting that these arguments and counterarguments also apply to modified genes, such as ESTs, CDNA, and genes attached to viral vectors. One could argue that an ESTs or genes attached to viral vectors are product of nature because they are the same as their naturally occurring counterparts.

57. For further discussion of pragmatic approaches to other metaphysical questions in biology, including questions concerning reductionism in biology (see note 8).

58. For further discussion of objectivity and values in science, Helen Longino, *Science as Social Knowledge* (Princeton: Princeton University Press, 1990); Philip Kitcher, *The Advancement of Science* (Oxford: Oxford University Press, 1994).

59. Steinbock, *Life Before Birth.*

60. Audrey Chapman, "Background and Overview," in *Perspectives on Gene Patenting*, ed. Audrey Chapman (Washington, D.C.: American Association for the Advancement of Science, 1999), 7–40.

61. Reginald Rhein, "Gene Patenting Crusade Moving from Church to Court," *Biotechnology Newswatch* (5 June 1995): 21; Evans, "The Uneven Playing Field on the Dialogue on Patenting."

62. Robert Merges, "Property Rights Theory and the Commons: The Case of Scientific Research," in *Scientific Innovation, Philosophy, and Public Policy*, ed. Ellen F. Paul, Fred D. Miller, and Jeff Paul. (New York: Cambridge University Press, 1996), 145–67.

63. Gold, *Body Parts.*

64. Michele Svatos, "Biotechnology and the Utilitarian Argument for Patents," in *Scientific Innovation, Philosophy, and Public Policy*, ed. Ellen F. Paul, Fred D. Miller, and Jeff Paul (New York: Cambridge University Press, 1996), 113–44.

65. Gold, *Body Parts*.

66. Evans, "The Uneven Playing Field on the Dialogue on Patenting," 57–74.

67. Amy Gutman and Dennis Thompson, *Democracy and Disagreement* (Cambridge, Mass.: Harvard University Press, 1996).

68. I see no a priori reason why the courts and the PTO are not qualified to address noneconomic values since these institutions rely on expert testimony in making their decisions. If they can review testimony from economic experts, then why can't the courts and the PTO also consider testimony pertaining to environmental, religious, moral, scientific, and other concerns?

eight

PATENTING GENES AND LIFE
Improper Commodification?
MARK J. HANSON

ontemporary American culture is suffused with market rhetoric. Virtually any feature of human experience is described in the language of an object that can be bought. Health, happiness, fertility, sex, physical traits, and even love are advertised as available for the right price. A few things long thought to be out of bounds are also now emerging as market possibilities on the near horizon, including human organs and babies with desired traits. At least in some places, the individual as a whole person escapes consideration as a commodity, even as his parts are being considered as fair game for the market.

Scientists' efforts to map the human genome are resulting in a remarkable picture of the links between genes and all sorts of traits as well as diseases. Within these genetic data lie possibilities for a range of medical-industrial applications, from genetic therapies for disease, to selection and enhancement of human traits, to manipulation of traits for multiple purposes. The gene for each application may be subject to a patent, an intellectual property claim that excludes those other than the patent holder from commercial exploitation of that gene without a license. Similarly, a genetically engineered nonhuman organism may also be patented, making it illegal for others to duplicate that organism without a license. Are these property claims going too far, morally? Do they represent an encroachment of the market into a place it should not be, namely, the traits of persons themselves and engineered life-forms?

In this vein, one of the central objections to patents on genes or genetically modified organisms is that such patents in fact do represent a commodification of their subject matter that is inappropriate. How should we understand this objection? In this chapter I will illuminate and analyze the components of this objection. It is my conclusion that patents on genes and genetically modified organisms represent a further encroachment of market thinking on subjective properties of persons and other living beings, but that this encroachment is highly attenuated and thus is not a sufficient cause for rejecting all patents. The commodification objection, however, is rooted in larger worldviews that raise important moral considerations that will be valuable to consider as the power of biotechnology increases, and as its commercialization grows.

PATENTS AND BIOTECHNOLOGY

A patent is an intellectual property claim on an invention, valid for twenty years. The subject matter of a patent is "any machine, article or manufacture, process, or composition of matter which meets the criteria of being new, useful, and nonobvious."[1] Although patents are a form of property rights and therefore ownership, the subject matter of a patent is not owned in the same way that one owns other things. Patents are a more limited set of rights than full ownership because they merely exclude others from making, using, or selling the subject matter for a limited period of time. To have a patent on a gene, for example, is not to own that gene as it exists in people's bodies. Patented genes are "isolated and purified" so as to be potentially used in some way.

Many issues regarding the patentability of genes are subject to legitimate debate, hinging on interpretations of the law—for example, how genes can be patented if objects of nature are not patentable, whether patenting actually constitutes ownership, or whether patents are ultimately harmful to medical practice. One thing is clear: patents do confer substantial power on the patent holder, especially the exclusive right to commercial exploitation.

Biotechnology is a multibillion-dollar industry, and it is growing rapidly, fed by venture capital and genetic data from the final months of the effort to map the human genome. By 1998, $97 billion had been invested in a U.S. biotechnology industry that consisted of 1,283 companies, employing more than 153,000 people.[2] According to an industry fact sheet, the largest five biotech companies spent an average of $121,400 per employee on research and development that year, compared with an average of $31,200 per employee for the largest pharmaceutical companies.[3]

For defenders of the current system, patents represent nothing more than a legal mechanism to ensure protection and just reward for the tremendous investment in research that may yield products of significant medical benefit. That a patent holder or licensee should also have the opportunity for profit is only suitable, given the investment not only of money but of considerable human resources as well. Patents therefore ensure an element of fairness and justice, encouraging and rewarding efforts while protecting the rights of those seeking to invent and develop useful items.[4] Further, it should be noted that the mere presence of a profit motive does not, in itself, morally taint the endeavor.[5] Typical of many industries in capitalist economic systems, altruism and self-interest coexist in an uneasy way.

The patent, however, also represents a basis for fierce competition for profits and commercial leadership. Companies are bought and sold for their patent rights. At the heart of the growing industry competition is a race by so-called gene hunters to patent genes so that they may be licensed for a fee to scientists and biotechnology and pharmaceutical companies. One company, Incyte Pharmaceuticals, literally ships genes to those seeking commercial exploitation of the genes, triggering a licensing agreement, and then collects royalties on those that ultimately result in product sales.[6]

Placing genes and genetically engineered organisms within the sphere of market competition has prompted the charge of commodification. For example, the race to patent the genes and cell lines of indigenous or homogeneous peoples has led to controversy over whether those people are being exploited as mere genetic resources, without proper consent or compensation. The creation and patenting of the so-called Harvard Onco-Mouse—a mouse that easily grows cancerous tumors—for some has turned the animal into a lifeless commodity, whose existence as a species is subject to limited monopoly control. In the well-known case of John Moore, his physician obtained a patent on a unique cell line derived from Moore's spleen without his consent. A long litigation process between patient and physician ensued.

In 1995 a coalition of more than two hundred religious leaders issued a "Joint Appeal against Human and Animal Patenting," two leaders of which defined "the commodification of life and the reduction of life to its commercial value and marketability" as a central motivation for their action. Motivated by similar concerns, the Council for Responsible Genetics, in 1996, sponsored the conference, "Resisting the Commercialization of Our Genes."

For these kinds of patents to generate such controversy, they must be more than mere legal mechanisms to protect investment. In fact, patents are not only the gateway to the marketplace, they are the gateway to different ways of understanding and valuing the subject matter of the patents. But why should this be a problem?

WHAT ARE GENES?

Whether understanding something in a new way is objectionable depends on what that thing is and how we currently understand it. We must therefore ask, What is it about our current understanding and valuing of the patentable subject matter that would make patenting of it in any way morally offensive?

How we understand many things is often variable. We can understand the same object in different ways simultaneously, such as when we give an insurance estimate on an object we have made but also consider it a priceless expression of ourselves. Or different people may understand something in various ways, such as a person understanding a human preembryo as mere biological tissue while others understand it as a full human person. Where is the truth in our debates on such matters?

The controversy about patents on human genes and genetically engineered organisms raises similar questions. We do not have long traditions of understanding how we should think of genes in relationship to our identities, as individuals or as a species. We have also not yet discovered all the relationships between genes and the traits to which they are linked. As the map of the human genome is completed, however, we are discovering daily the integral role genes play in human traits of all kinds, behavioral as well as physical. We are also discovering that genes alone do not define who we are. Environmental influences—and, for some, factors of mind and spirit—also play a role.

Knowing these facts is a beginning to our understanding of genes' biological reality. But because genes are implicated in personal and species identity, how we understand them will necessarily be implicated in larger worldviews according to which people understand their identity. Therefore, there is no definitive biological or legal definition of a gene that will

resolve these debates once and for all. Debates about the moral and philosophical significance of gene patenting cannot, therefore, be merely a
matter of the "facts" about the legal status of patents. This is not to say that
worldviews are not immune to criticism—just as people may legitimately
debate whether interpretations of religious doctrine within a tradition are
correct. But neither can worldviews be dismissed in favor of "neutral" or
"objective" notions of reality, for all such notions are embedded and intelligible within larger worldviews of one kind or another.

Consider how DNA may be variously understood. One way of
understanding DNA fits very well into a dominant language of contemporary culture, namely, DNA as information. Richard Dawkins's view
reflects the language of today's "information age."

> Life is just bytes and bytes and bytes of digital information. Genes are
> pure information—information that can be encoded, recoded, and
> decoded, without any degradation or change of meaning. Pure informa-
> tion can be copied and, since it is digital information, the fidelity of the
> copying can be immense. . . . We—and that means all living things—are
> survival machines programmed to propagate the digital database that did
> the programming.[7]

This perspective contrasts sharply with a view of human DNA put forward by Richard Land and C. Ben Mitchell of the Southern Baptist Convention's Christian Life Commission: "Admittedly, a single human gene or
cell line is not a human being; but a human gene or cell is undeniably human
and warrants different treatment than all non-human genes or cell lines. The
image of God pervades human life in all of its parts."[8] For Land and
Mitchell, DNA is sacred, inseparable in value from the image of the divine.

Both of these perspectives are rooted in larger worldviews and carry
with them strong implications of what is appropriate behavior vis-à-vis
genetic material. To "recode" information "without degradation or change
of meaning" is to suggest that any DNA manipulation is unproblematic.
To call DNA sacred is to limit severely what can be done to it.

A third perspective is implicit in the view that genes are patentable
subject matter. In legal parlance, genes are "mere chemicals." To patent
something is, first of all, by statute, to describe it as a "machine, article of
manufacture, process, or composition of matter." It is also to describe it as

an appropriate subject for commercial exploitation. Whatever is patentable becomes subject to a new way of understanding it, namely, in the context of the free market. Furthermore, to codify this description in federal regulations is to place it into a broader cultural lexicon of understanding. In short, to call DNA patentable is to call it marketable. It is the tension set up by the juxtaposition of this understanding and those embedded in other worldviews that is the root of the commodification objection.

UNDERSTANDING COMMODIFICATION

To understand this issue more fully, it is helpful to consider the notions of commodity and commodification themselves. A commodity is an economically valued good within a relationship of exchange.[9] The commodity's value is based on its potential for exchange in a process otherwise known as consumption.

The notion of commodification, however, is more complex than market consumption. Law professor Margaret Jane Radin clarifies this by defining commodification as "the *social process* by which something comes to be apprehended as a commodity, as well as to the state of affairs once that process has taken place."[10] By this definition, commodification becomes a rather complex idea, admitting of degrees and depending on social context. According to Radin, commodification may be incomplete, a condition that exists when market and nonmarket understandings coexist, or when only certain segments of a society (but not others) understand something as a commodity.[11] Analyzing commodification therefore requires investigating a continuum of social understanding.

In addition, commodification can, according to Radin, be understood in narrow and broad senses. The narrow sense of commodification applies when things are literally bought and sold in the marketplace; the broad sense applies when things are discussed in market terms, even if they are not bought and sold.[12] The object of market discourse can be virtually anything, even, in the present case, human traits: "In market rhetoric, the discourse of commodification, one conceives of human attributes (properties of persons) as fungible owned objects (the property of persons)."[13]

How does patenting involve commodification? The role of the patent in the commercial exploitation of genetic material certainly involves commodification in the broad sense. Without a motive to control and profit from commercial development of a product, patents would be of little use. Patents, for all practical purposes, signal a commercial interest of some kind, and most often a profit motive. In short, applying the language of patents to genes is to add the language of the market to human traits.

Although patents on genes and genetically engineered organisms are not as comprehensive as full ownership rights, they imply enough sovereignty over the disposition of the patented material that literal buying and selling is facilitated. A patent does not, for example, constitute an ownership claim on genes as they exist in a person's body. Thus, a patent holder on that gene does not "own" a part of you. The holder can, however, make money from copies of that part of you. That Incyte Pharmaceuticals literally ships copies of genes to companies—from their "clone by phone" factory—and collects royalties on the profits from the products developed from those genes demonstrates an encroachment of commodification approaching even the narrow sense of the term.[14]

A MORAL PERSPECTIVE ON THE COMMODIFICATION OF GENES

But what is wrong with this, especially if those companies are often using these genes to create products that will be beneficial to human well-being? Commodification, in narrow or broad sense, is not—according to most worldviews—in itself wrong. In fact, Americans accept at least an incomplete commodification of almost everything. But at the heart of the commodification objection to gene patents is the conviction that these patents represent an encroachment of market discourse and ways of valuing properties (that is, traits) of persons. This step is perilously close to the worldview in which the basic components of life are seen as appropriate subjects of human manipulation and even profit. This is something that people have often held to be morally troublesome.

The science of genetics is beginning to allow us to conceive of traits of persons in ways that are increasingly objectifiable. With growing frequency, the media report links between genes and human traits of all kinds. Some of these traits are disease traits, while others are physical features or traits of behavior, personality, intelligence, and even character. To identify a link is to give a trait a biological locus that is isolable and can literally be seen under the microscope. And the more we identify links between genes and traits, the more we invest our identity itself in our genes.

There is a cultural and philosophical tradition in the West, rooted in Kantian philosophy, that recognizes a distinction between the self and its properties as a subject, and the nonself as objects. While contemporary philosophy is whittling away at so neat a distinction, we nevertheless still often find morally objectionable the objectification of persons and their attributes.[15] Opposition to slavery and prostitution are obvious examples of this common to many moral traditions. Persons, being subjects, are to be treated as ends in themselves, and never to be treated merely as means, that is, subjects of manipulation and control. Objects, on the other hand, can be treated as mere means. Turning a trait of the self into a commodity, on this line of thought, objectifies part of the self by treating it as a mere means.

It is on the basis of this kind of thinking that philosopher Baruch Brody posits the maxim that "it is wrong to commercialize something with which individuality and personhood is intertwined."[16] His conclusion is that commodification objections would only prohibit patents on the entire genome of an individual person. I believe this conclusion is correct, but that the issue is more complex than this, hinging on the degree to which we identify ourselves with our genes.

I have argued elsewhere that we are best served by avoiding two extremes: a genetic reductionism that equates persons with genes and a materialism that "empties" the genes of any significance whatsoever regarding our individuality and personhood.[17] The former view not only neglects the external influences that shape human identity, it neglects the complex ways by which we come to define ourselves as unique persons. The latter view neglects the central role that even fractions of our genomes can play in defining us both as human and as individuals. To acknowledge this as true is to accept that something morally significant is at stake not only when we manipulate genes, but also even as we talk and think about them.

On this view, the offense of commodification hinges on how we understand genes and their relationship to human personhood. It also hinges on what exactly is being patented. Patents on generic human traits would be offensive in a way that those on disease mutations, which are not integral to human identity as such, are not. In fact, people experience illness as an alienation of the self from the body in which the body becomes objectified.[18] Patents on disease genes are less morally offensive because they are on aspects of ourselves that we already experience as alienating. Diseases don't belong to us in the same way our other self-constituting traits do. They are not who we are. Rather, they reinforce self-alienation. It, therefore, is less troublesome to moral intuitions that biotechnology companies patent disease-related traits, but more troublesome that they seek patents on human genes in a wholesale fashion.

The moral picture I am developing here is one in which the objection to human gene patenting is highly qualified. If we accept that patents are a form of market discourse and that genes are integral to human personhood and individual and species identity, commodification occurs to a degree through the application of patents to human traits. This commodification is, however, both broad and incomplete in Radin's terms. It is broad in that patenting in itself does not directly entail the buying and selling of genes, and it is incomplete in that it does not exclude other ways of valuing the human person and is not universally accepted. Despite the manner in which the commodification objection must be attenuated, there are legitimate concerns to consider. I will do so after a brief discussion of patents on living organisms.

PATENTS AND THE COMMODIFICATION OF LIVING ORGANISMS

The patenting controversy began with the *Diamond* v. *Chakrabarty* decision in 1980 by the U.S. Supreme Court that a genetically modified microorganism could be patented. Since that decision, many modified organisms have been patented. Commodification of these organisms, however, is not thought to be problematic in the same way that commodification of human beings and their traits is. Plants and animals have been owned, bought, and sold for centuries. We have bred them intentionally to alter their genetic constitutions to suit human desires for companionship, showmanship, agriculture, commerce, and medical research.

The *Chakrabarty* decision was notable, however, in that the subject matter's status as living was irrelevant to its patentability. Philosopher Leon Kass argues that in the decision, the Court promoted a new dimension in how all organisms are understood.

[I]n [the Court's] eagerness to serve innovation, it has, perhaps unwittingly, become the teacher of philosophical materialism—the view that all forms are but accidents of underlying matter, that matter is what truly is—and therewith, the teacher also of the homogeneity of the given world, and, at least in principle, of the absence of any special dignity in all of living nature, our own included.[19]

On this view, patents would promote a more complete form of commodification by ignoring any value to an organism based on its status as living. The sense of commodification and ownership is further expanded beyond merely the possession of individual organisms. A patent entails a limited property right monopoly over entire newly engineered species. Kass argues, "It is one thing to own *a* mule; it is another to own *mule*. Admittedly, bacteria are far away from mules. But the principles invoked, the reasoning, and stance toward nature go all the way to mules, and beyond."[20]

This statement hints that the central concern here is primarily with what patenting life-forms may mean for shifting our worldviews toward new perspectives of nature that minimize respect for organisms as living and that tend toward views of organisms as objects of human manufacture. Kass provocatively concludes, "If a genetically engineered organism may be owned *because* it was genetically engineered, what would we conclude about a genetically altered or engineered human being?"[21]

Of course, in practical terms, the patentability or ownership even of genetically engineered human beings would be more obviously morally offensive than patents on animals or individual human traits. It would also not be legally permitted. But Kass's point again indicates the degree to which decisions about patentability may be benchmarks in an ongoing evolution of worldviews.

RHETORIC OR REALITY?

It may be objected at this point that commodification primarily through mere market rhetoric—especially one as seemingly esoteric as the language of patent law—ultimately pales in moral significance relative to the advances in human well-being that may be facilitated by research supported by the current patent system. This may be true on its face, although the commodification objection is also one among many possible objections to patents for biotechnology products.

To be concerned about the commodification entailed by patents is to be concerned about how new ways of understanding something can advance certain worldviews and values, and in the process diminish other worldviews and values. Those objecting to patents on genes and genetically engineered organisms on the grounds of commodification may not be offering Kantian-type philosophical arguments about subject and objects, or even very sophisticated theological accounts of the sacredness of life or human biological materials. They are concerned, however, about a reductionism of that which possesses some kind of intrinsic value to economic common denominators and the advancement of a worldview that easily facilitates such a move by only considering economic values.

Even worries about partial commodification in rhetoric are ultimately worries about further encroachment to more literal, complete, and inappropriate forms. Radin summarizes the broader point well in relation to the commodification of individual traits.

> Systematically conceiving of personal attributes as fungible objects is threatening to personhood because it detaches from the person that which is integral to the person. . . . Moreover, if my bodily integrity is an integral personal attribute, not a detachable object, then hypothetically valuing my bodily integrity in terms of money is not far removed from valuing me in terms of money.[22]

The rhetoric of the market or other worldviews matters because it is not merely a way in which we describe reality; it also instantiates our values and defines appropriate ways of behaving. Legal rhetoric is especially important because it is a socially common way of describing and prescribing appropriate behavior. It promotes certain views of ourselves and relations to others. Law professor Cathleen Kaveny makes the point more starkly: "Legal instantiation changes the thing itself."[23]

Defenders of gene patents have argued that objectors are not concerned so much about patents as they are about abuse or immoral practices in biotechnology, practices that for the most part do not exist. My analysis, however, demonstrates that the link between former and latter lies in how patents facilitate a worldview in which certain objectionable practices are more likely because of the way in which human genetics and engineered organisms are commercialized. There is a worry, therefore, that subtle shifts in understanding facilitate the ease of genetic manipulation and abuse. As a rhetoric of the market and property rights, patents are a rhetoric of ownership, control, and assertions of sovereignty. The key here is understanding the boundaries of the link between the rhetoric of patents and the actions of scientists and biotechnology companies. Rhetoric, therefore, matters morally; it is never mere rhetoric.

MORAL IMPLICATIONS

If commodification can admit of degrees and be described as a process as well as a state of affairs, it becomes more difficult to draw firm moral and especially policy decisions about an issue based on commodification alone. On the issue of patenting genes and genetically engineered organisms, I have argued that patents represent an *encroachment* of commodification on our understanding of traits and organisms. The attenuated nature of this encroachment (being broad and incomplete), however, makes drawing out policy implications even more difficult.

If we recognize the problematic nature of applying market rhetoric to properties of persons as a process, obligations can be recognized both to limit the degrees of commodification and also, more important, to inhibit the kinds of behaviors and views of persons and nature that commodification promotes. For example, attempts to profit from human beings as genetic resources, such as the patenting and pursuit of profit from the cell line of John Moore's spleen or from the genes of homogeneous populations without their consent, display behaviors that instantiate a worldview of complete commodification—people and their traits treated as commodities rather than as significant features of persons. Recognizing obligations to informed consent with accompanying dialogue regarding the worldviews that shape understandings of genes or patentable natural substances mitigates a more complete commodification of persons.

Recognizing moral, philosophical, and religious ways of valuing genes and organisms also helps to militate against the ease in which the commodified can be easily manipulated for purposes of profit and control. Given what is at stake in such manipulations, society has a moral interest in maintaining plural ways of understanding such matters and guarding against more complete commodification. The commodification objection should also prompt patent policymakers to consider limiting patentability to only certain kinds of biological materials, namely those with less direct connection to our personhood and personal identity.

The morality of patenting genes and organisms hinges on a greater complexity of issues than commodification alone. The benefits of the U.S. patent system are easily demonstrated. Assertions of those with the greatest financial interest notwithstanding, however, there are great enough reasons that alternatives to the current system should be explored.

Ultimately, some form of commodification of life-forms and their traits will be inevitable, as long as medical research and practice are governed largely by the free market. The commodification objection to patents, regardless of its weakness, nevertheless serves an important moral function. It reminds us of the tensions between our motives to serve and our motives to capitalize on the needs of others. It reminds us of the tensions between the biological and nonbiological components of our identities. And it reminds us of the importance of maintaining plurality in how we understand ourselves and the living world around us. These reminders will become increasingly morally valuable as we assume greater and greater powers over life itself.

NOTES

1. Kate H. Murashige, "Intellectual Property and Genetic Testing," in *The Genetic Frontier: Ethics, Law, and Policy*, eds. Mark Frankel and A. H. Teich (Washington, D.C.: American Association for the Advancement of Science, 1994), 183.

2. From the Web site of the Biotechnology Industry Organization (http://www.bio.org/news/stats.asp).

3. Ibid.

4. Leon R. Kass, *Toward a More Natural Science: Biology and Human Affairs* (New York: Free Press, 1985), 135.

5. J. W. Child, "Profit: The Concept and Its Moral Features," *Social Philosophy & Policy* 15, no. 2 (1998): 243–82.

6. "Gene Quest Will Bring Glory to Some: Incyte Will Stick with Cash," *Wall Street Journal*, 10 February 2000, A1.

7. Richard Dawkins, *River Out of Eden: A Darwinian View of Life* (New York: BasicBooks, 1995), 19.

8. Richard D. Land and C. Ben Mitchell, "Patenting Life: No," *First Things* 63 (May 1996): 21.

9. Mark J. Hanson, "Biotechnology and Commodification within Health Care," *Journal of Medicine and Philosophy* 24, no. 3 (1999): 268.

10. Margaret Jane Radin, *Contested Commodities* (Cambridge, Mass.: Harvard University Press, 1996), xi (emphasis added).

11. Radin, *Contested Commodities*, 102–103.

12. Ibid., 12.

13. Ibid., 13.

14. "Gene Quest," *Wall Street Journal*, A10.

15. Radin, *Contested Commodities*, 155–56.

16. Baruch Brody, "Protecting Human Dignity and the Patenting of Human Genes," in *Perspectives on Gene Patenting: Science Religion, Industry, and Government in Dialogue*, ed. Audrey R. Chapman (Washington, D.C.: American Association for the Advancement of Science, 1999), 118; see also Margaret Jane Radin, "Residential Rent Control," *Philosophy and Public Affairs* 15 (1986): 350–80.

17. Hanson, "Biotechnology and Commodification," 277.

18. Kaye Toombs, *The Meaning of Illness: A Phenomenological Account of the Different Perspectives of Physician and Patient* (Dordrett: Kluwer, 1993), 90.

19. Kass, *Toward a More Natural Science*, 149–50.

20. Ibid., 151.

21. Ibid., 151.

22. Radin, *Contested Commodities*, 88.

23. Cathleen Kaveny, "Religious Values, Cultural Symbols, and the Law." Paper presented to the Religion and Biotechnology Project at The Hastings Center, Briarcliff Manor, New York, 18 April 1997.

nine

FALLING FROM GRACE
Science and the Pursuit of Profit
ROBERT LEE HOTZ

As it unfolded over five weeks in a San Francisco federal courtroom during May of 1999, the bitter dispute over the billion-dollar molecule turned a chronicle of scientific breakthrough and enviable business reward into pulp fiction.

Expert testimony about the arcane art of molecular biology and the legal niceties of rival patent claims was overshadowed by lurid allegations of cocaine abuse, commercial pressure, laboratory intrigue, divided loyalties, falsified research, fraud, and a mad scramble for spoils.[1] There was even a startling courtroom confession of a surreptitious New Year's Eve foray by company employees into a university lab to acquire critical samples of the lucrative gene.[2]

So went the trial of the *University of California* v. *Genentech Inc.*, which pitted the educational institution where the genetic engineering revolution first flowered against one of the first and arguably most successful of the biotechnology companies founded to capitalize on that revolution. The molecule in question was the genetic sequence that encodes the protein for somatotropin, the human growth hormone used to treat some 100,000 children worldwide. With cumulative sales of the company's biosynthesized hormone totaling $2 billion between 1985 and 1999, that single snip of DNA was the foundation of Genentech's prosperity.[3]

The university sought $400 million in back royalties, and, if a jury agreed that the patented gene had indeed been willfully misappropriated by Genentech, the company in addition would be liable for triple damages— about $1.2 billion in all.[4] The jury deadlocked. Within months, Genentech settled the case by paying the University of California $200 million, while the scientist involved faced sanctions for knowingly publishing false data.[5]

The dispute is a cautionary tale of the turbulent forces affecting scientists in the brave new world of corporate biotechnology, in which the powerful drive for commercial profit has added new ethical and moral challenges for scientists seeking to harvest the genomes of plants, animals, and human beings. The lawsuit, and others like it, are symptoms of fundamental forces that affect how openly many researchers conduct their experiments, who they work for, the questions they seek to answer, how freely they share their results, and how honestly they talk about what they do.

From its beginnings in basic breakthroughs during the 1970s in the ability to manipulate directly the molecules of heredity, the practice of modern molecular biology has been accompanied by profound misgivings about how such knowledge might be misused.

Leading biologists were especially worried about how the rapid commercialization of such intimate manipulations of life might affect the behavior of life scientists. There were concerns that increased dependence on commercial funding would undermine the integrity of university research, affect the public research agenda, delay publication of results, stifle informed scientific dissent, and interfere with the conduct of research on campuses—especially as scientists themselves eagerly launched their own entrepreneurial ventures. There is substantial evidence that those early misgivings were well founded.

In this chapter, I intend to review some of the pressures operating on life scientists today and to examine how the growing commercial involvement in university research has undermined some traditional scientific practices. Clearly, the relationship between life sciences companies and university scientists is an evolving one that has prompted a healthy, albeit sometimes painful, reassessment of the common research practices and courtesies on which scientific progress so often depends.

Certainly, when Congress in the early 1980s lowered the barriers between industry and publicly funded research universities, its goal was to encourage innovation, to bolster the competitive strength of U.S. business, and to speed the dissemination of new research into the markets where it might most quickly benefit the public. Changes in federal technology transfer regulations fostered more intimate relationships between publicly financed laboratories and commercial biology ventures. And federal researchers were encouraged to patent the fruits of their research, to profit directly from their discoveries.

Between 1991 and 1996, the academic share of new patents doubled.[6] Indeed, 40 percent of all the patents on human DNA sequences issued in 1995 went to universities, double the estimate for the previous decade.[7] More than one hundred universities launched their own for-profit ventures. In all, the commercialization of academic research led to some $33.5 billion in economic activity and 280,000 jobs in 1998, according to the Association of University Technology Managers, which represents three hundred U.S. research universities.[8]

In the resulting gray areas between academia and industry, the struggle for scientific truth also became a market struggle for commercial advantage—a contest played under different rules, with different incentives and penalties than those of traditional science. For researchers in medicine and the other life sciences, schooled in one professional ethic and expected to thrive under an entirely different set of values, the problems were most obvious and, perhaps, most troubling.

"The increasing involvement of for-profit biotechnology in medical research has provided new sources of funding, but with their involvement has come an emphasis on the ethical and operational values of business rather than science," said pioneering oncologist Steven A. Rosenberg, chief of surgery at the National Cancer Institute.[9]

University scientists who might otherwise be colleagues in the scientific enterprise became business competitors jockeying for commercial advantage. Moreover, there soon was a greater willingness to suppress or distort research, as several peer-reviewed studies have shown.[10] Scientists became less willing to share critical research tools and biomaterials.[11] Old habits of openness and collegiality gave way to nondisclosure agreements and material transfer licenses.

In the field of medical genetics, commercial stakes were so immense that, in the most extreme cases, even a well-regarded scientist could be tempted to lie and steal, as testimony in the *Genentech* trial demonstrated, to bolster a failing commercial research effort. In a similar case, a federal judge in San Francisco ruled in December 1999 that scientists from the Cetus Corporation, which is now owned by Hoffman-La Roche, intentionally misled the U.S. Patent and Trademark Office to obtain a patent on the naturally occurring form of the reagent Taq, which is the most purchased reagent under National Institutes of Health (NIH) grants.[12] In his ruling, Judge Vaughn Walker likened the researchers' behavior to scientific misconduct.

Indeed, so willing were some new businessmen of biology to overturn traditional checks and balances in their struggles for exclusive control of potentially lucrative molecules that even the validity of peer review itself has gone on trial. In a recent trade dispute between Cistron Biotechnology and the Immunex Corp., lawyers marshaled experts in a Seattle courtroom to argue that the academic idea of confidential peer review simply has no legal validity and that any scientist should feel free to lift interesting data from a research paper sent to him for peer review.[13] At issue were patent rights to a potentially valuable protein gene.

The litigation eventually settled out of court.[14] The molecule turned out to be worthless. But the case sent a chill through the scientific community over the basic question of mutual trust that underpins peer review and the formal assessment of any new research.

"There is no question that the culture of the biological sciences has changed," said Sheldon Krimsky, a biotechnology analyst at Tufts University in Medford, Massachusetts, who has closely monitored the changing relationships between industry and academia over the past thirty years.[15] "We now have other values we have to consider. We are no longer just interested in pursuing knowledge. We are also trying to market it."

Grove City College president John H. Moore, an economist who is a former president of Sigma Xi, a research society that includes 80,000 scientists and engineers as members, said that as corporate funding of the life sciences increases, the profit motive is unavoidable—and almost irresistible. "In the biological sciences, people see a tremendous pot of gold at the end of the rainbow and they are much more tempted to get into these arrangements than, say, an astronomer," Moore said.[16] "There is not much money to be made from discovering a new black hole."

To be sure, much more than the impulse for personal profit is at work. Several other factors have combined to intensify the professional and financial pressures on scientists today, making commercial relationships increasingly attractive for university-based researchers.

Money, like insight and imagination, is the lifeblood of research. Since 1960, patterns of research and development (R&D) funding have changed dramatically, increasing the commercial influence over research in the United States. In 1960 about 65 percent of all R&D funds were provided by the federal government and only 33 percent by industry. By 1995 the proportions had almost reversed.[17]

During those years, the science of life created a major industry, with more than 1,300 companies, nearly $13 billion in annual revenues, and more than 100,000 people on its payroll.[18]

It transformed the practice of medicine. Tens of millions of people have used genetically engineered medicine to control heart disease and diabetes, treat cancer, or offset the effects of stroke, according to the Biotechnology Industry Association. Tens of thousands more routinely undergo genetic testing—either to judge the health of their unborn children or to tell if they themselves are susceptible to chronic diseases or cancer. In 1995 alone, medical researchers patented 500 genetic engineering products. And the pace of development was accelerating. Industry surveys show that more than 284 new genetically engineered medicines for diseases ranging from the common cold to AIDS were tested in 1995—up almost 20 percent from the year before.[19]

In addition, every year farmers sow tens of millions of acres of U.S. farmland with the patented seeds of genetically engineered crops—and are being sued for seed piracy if they use seeds from their harvest for replanting the following season. Between 25 percent and 45 percent of some major U.S. crops already are genetically modified. Industry officials expect 90 percent of U.S. agricultural exports to be genetically engineered within a decade.[20]

Instead of spawning rogue microbes or ecological disasters, the industry instead has sharply changed ideas about the ownership of living organisms while researchers have raced to patent the creatures they create and the genes they find.

For example, scientists at Case Western Reserve University who recently discovered how to create artificial human chromosomes conducted most of their research with public funding from the National Institutes of Health. They patented the technique and acquired a corporate partner to exploit it before the research ever became public.[21] Physicians who treat families with genetic diseases are approaching geneticists and offering to "sell you my families"—meaning that they will, for a fee, give the researcher their patients' blood samples.[22] Scientists who isolated certain genes from the blood samples then patent them and profit from their use in genetic tests. Since the mid-1980s, when the U.S. Patent and Trademark Office started granting patent rights for human genes, it has received over 5,000 patent applications and has granted more than 1,500. More than fifty biotechnology companies are developing or providing tests to diagnose genetic disorders or to predict the risk of their future occurrence.

Such a thorough mingling of public research and private enterprise, although troubling to some, is now the rule. "The tens and tens of billions of dollars from biotechnology companies have fundamentally changed the way science is done in the United States," said Stanford University anthropologist Paul Rabinow, who studies the business of molecular biology and genetics.[23]

Looking solely at the financial relationship between universities and life sciences companies, the commercial influence had become pervasive. By 1994, 90 percent of companies involved in the life sciences had developed formal ties with academic researchers, and 92 percent of university-based researchers in the life sciences reported they received some form of support from companies. In all, the companies funded an estimated 6,000 academic research projects at a total cost of about $1.5 billion—about one of every ten research dollars spent by universities that year.[24] By some estimates, the corporate sponsorship of university-based medical research grew from about 5 percent in the early 1980s to as much as 25 percent at some schools by 1999.[25]

While commercial funding continued to be dwarfed by federal research spending, it assumed a greater importance as the number of new scientists grew faster than public spending on research. In a glut that began building in 1987, universities were producing about 25 percent more doctorates in science and engineering fields than the economy could easily absorb.[26] The number of Ph.D.s holding full-time research positions five years after graduation decreased from 89 percent in 1973 to 62 percent in 1995. Indeed, the raw competition between scientists was becoming "destructive," especially in the life sciences, the National Research Council warned.[27]

Even as the public funding for life sciences steadily increased, it was not enough to keep pace even with mild inflation or with the staggering growth in the size of the research community itself. By 1998 officials at the National Institutes of Health approved one of four worthwhile grant proposals they received, compared to one in three a decade before.[28] Consequently, the fierce competition for research support drove many scientists to seek industry funding—with its trade secrecy and market competition—more aggressively.

At the same time, many biotechnology companies discovered that their own corporate fortunes were equally dependent on the influx of new ideas, new techniques and, ultimately, marketable products from the academic laboratories they supported.

In a 1996 survey of 210 life sciences companies, David Blumenthal and his colleagues at the Health Policy Research Unit at Massachusetts General Hospital found that companies were significantly more likely to be supporting academic research than a decade earlier.[29] Almost two-thirds of the firms reported that their research funding had yielded patents, products, and sales. A third said that they depended on faculty to invent products the company could license.

While there was nothing new about commercial research in a university setting, these partnerships in biology were more varied, aggressive, and controversial. When Sheldon Krimsky, at Tufts, studied the ties between university biologists and 291 genetic engineering firms in 1984, he found hundreds of academic scientists already serving as consultants or as corporate officers.[30] He also found that two-thirds of the scientists with such commercial ties also reviewed grant proposals for the National Science Foundation. And nearly one in four of the scientists working on biotechnology-related sections of the National Academy of Sciences also served on the board of such a firm.

As such relationships became even more intertwined, it was increasingly difficult to tell just whose interests a university scientist might be serving. In fall 1998, for example, plant geneticists at University of California at Berkeley signed a five-year, $25-million research contract with a Swiss-based life sciences company called Novartis.[31] All but three scientists in Berkeley's thirty-two-member Department of Plant and Microbial Biology signed up to participate in the deal. Although the public university has strict rules on publication of research and will retain ownership of the research results and patents, the contract stirred concerns on campus and in the state legislature at the prospect that one company could have such influence over an entire department and the potential impact on academic freedom.

Writing in the *California Monthly*, the university's alumni magazine, Berkeley law professor Robert Berring said, "We must ask at what point does the university bargain away so much of itself that it ceases to be the university and becomes a partner of the private sector? If the private sector begins to play a role in setting the research priorities, if it gets built into the budget, something very fundamental has changed."[32]

Dominating the courtyard of the National Academy of Sciences in

Washington, D.C. is a statue of Albert Einstein bent in contemplation of an inward universe—a figure at once both childlike and monumental. The inscription on its pedestal is a succinct expression of a basic tenet of science:"The right to search for truth implies also a duty: One must not conceal any part of what one has recognized to be true."

In science, a secret not only impedes the general advance of new knowledge, it also distorts the understanding of what is generally held to be true about the world around us by withholding information that might force scientists to reinterpret what they already have learned.

Secrecy, however, is often the price of biomedical research funding. Corporate donors and the academic community increasingly are at odds over the release of findings and the sharing of critical research materials. "The commercialization of science has led to a new regimen of secrecy that is of great concern to the scientific community . . . secrecy of an entirely new scope and scale," said physicist Irving A. Lerch, who spearheaded an effort by the American Association for the Advancement of Science to combat the trend.[33] "It is very troubling."

Certainly, many scientists routinely keep new research to themselves, however briefly, to preserve a fleeting competitive advantage in the academic race to publish or perish, to be the first to announce a new discovery, or to secure foreign patent rights. Oil company geologists keep secret their stores of data about potential petroleum reserves, just as corporate chemists dutifully guard a proprietary soft drink formula, an over-the-counter nostrum, or a new shampoo. Medical researchers protect the privacy of their research subjects.

"Secrecy is not always bad," said John Deutch, a former director of the Central Intelligence Agency who now teaches at the Massachusetts Institute of Technology.[34] "But it is a major threat to science when it arises and it is antithetical to the purposes of a university."

To be sure, science has always depended on ground rules set by its patrons, ranging from private foundations and industrialists to the federal agencies that have integrated science into virtually every aspect of national life. As pharmaceutical companies and other biomedical endeavors assumed an ever increasing share of university research budgets, however, commercial secrecy became more common on campuses.

Companies often wanted to withhold research findings, experimental materials like cell lines or genetically engineered lab animals, and research methods they underwrite. Their motivation was to wring maximum advantage from their investment, protect a market share, or simply to deny information to a potential competitor.

Patent attorneys also regularly advised researchers to limit what they tell their colleagues about their work before they apply for a patent because patent law requires that the content of the patent not be revealed in "prior art." Oral reports, abstracts, grant proposals, and published papers all are considered to be prior art.

So, for example, Ian Wilmut and Keith Campbell at the Roslin Institute in Scotland did not announce the cloning of the first adult mammal, the white-faced sheep dubbed Dolly, in 1995 until the institute had applied for patents on cloned sheep.[35] In the same vein, a number of private life sciences companies in 1999 refused to share crucial genetic sequence information with public health authorities and federal researchers on at least six lethal microbes—among them *Staphylococcus aureus*, the most common source of life-threatening bacterial infections in hospitals; *Mycobacterium tuberculosis*, which causes TB; and *Enterococcus faecalis*, second only to staph has a cause of potentially life-threatening hospital infections—pending resolution of their various patent claims.[36]

Withholding DNA sequence data on pathogens could cost human lives, noted *Journal of the American Medical Association* (*JAMA*) West Coast editor Drummond Rennie, but nonetheless it is "commonplace." In 1995, for example, biotechnology companies withheld the entire genetic blueprint of a bacteria linked to stomach ulcers and gastric cancer—*Heliobacter pylori*—to keep the information from aiding rivals.

"The resulting undermining and reversal of the biomedical culture of open communication and exchange are among the most destructive impacts of life patents," wrote Jonathan King, professor of molecular biology at the Massachusetts Institute of Technology, and Doreen Stabinsky, an assistant professor of environmental studies at California State University at Sacramento.[37] Both serve on the board of directors of the Council for Responsible Genetics, in Cambridge, Massachusetts.

It could be equally disruptive when, for its own financial purposes, a company decided to prematurely disclose proprietary research. Many scientists were dismayed, for example, when in July 1999 a biotechnology company called DiaSorin bypassed traditional channels of scientific communication to reveal that its researchers had identified a new hepatitis virus in a corporate press release, rather than submitting the findings to a peer-reviewed journal or discussing the discovery at a scientific meeting.[38] The news was timed to coincide with an announcement of the firm's second-quarter earnings. Ironically, the scientist responsible for the discovery had not prepared a detailed scientific report because he had been warned by company lawyers to hold off for at least a year in order to avoid jeopardizing any patent claims involving the virus.

In a variety of such ways, scientists had lost control of their work. Many researchers found themselves in a world in which intellectual property rights, material transfer agreements, "reach-through" clauses, and nondisclosure forms trumped professional ethics and academic freedom.

As a consequence, scientists who once routinely shared new research tools for experiments now found they had to negotiate a thicket of restrictive licensing agreements to get access to new generations of genetically engineered "knockout" mice that serve as laboratory models of human diseases, gene fragments called expressed sequence tags (ESTs), reagents, safe dosage–level data, gene amplification techniques, and equipment. The licensing requirements for Du Pont's patented Cre-loxP mice so rankled scientists that they drew protests from National Institutes of Health (NIH) director Harold Varmus, who appointed a panel to determine whether the restrictions would impede research.[39] Eventually Varmus secured an agreement that eliminated the most onerous limitations on research materials.

"The problem these days is that commercial organizations try to attach very severe restrictions on our ability to function in an open manner," said Nobel laureate molecular biologist David Baltimore, president of Caltech.[40] "It is a problem in my own laboratory. . . . If the laboratory ever lost its sense of openness, we would lose one of the core values that lead to great science," Baltimore said. "Openness is a necessity."

But such openness could no longer be taken for granted. Growing numbers of researchers have financial entanglements, and many face restrictions on what their employers allow them to talk about openly— rules that greatly hinder the free exchange of scientific information.

In the first large study of collaborations between academia and industry, researchers at Carnegie Mellon University studied 1,058 university-industry research centers at 203 universities and found that a third of the agreements allowed the industrial sponsor to suppress research.[41] The center agreements affected 15 percent of the nation's academic science and engineering workforce—some 12,000 faculty and 22,300 other doctoral-level researchers. Total spending of all the centers was $4.29 billion in 1990—the year under review—more than twice the research budget of the National Science Foundation.

Half of the sponsorship agreements allowed the company to delay publication. Most of the firms also reported that they sought to keep research secret as long as possible, often up to six months or more—three times longer than NIH guidelines or many university guidelines would consider reasonable.

More than a third said they had the right to restrict communication about their research with the public. And almost all of them said they required students and postdoctoral fellows to keep secret any proprietary information that arose during an experiment.

Other financial ties are harder to detect and may elude university or journal disclosure policies. A team of Harvard researchers looked at the questions of gifts, and, in a study published in *JAMA* in April 1998, they questioned more than 2,000 scientists to find out how often researchers in the life sciences had been offered presents from companies. They found that 43 percent of the scientists had been offered presents from companies in the form of useful biomaterials, equipment, or travel support, and almost a third said the donors wanted prepublication review of any articles stemming from use of their gifts.

To determine how often scientists delayed publication of their research or refused to share results with colleagues, health policy analyst David Blumenthal and his colleagues at Mass General recently surveyed 3,400 scientists at the top fifty research universities that receive the most taxpayer support from the NIH.[42] One researcher in five admitted delaying publication for more than six months one or more times in the past three years to protect a scientific advantage, patent a discovery, or secure some other commercial advantage. Of those, one in four admitted withholding information simply to slow dissemination of undesired results. One in ten had flatly refused to share research results with other university scientists at least once in the past three years. By the same token, more than a third of the scientists Blumenthal queried said they had been denied access to research results or lab materials by other university scientists.

In general, those who received commercial research support were the most likely to be closemouthed about their work. They were twice as likely to delay research publications as those who did not receive commercial support. They also were more likely to deny other scientists access to research results or lab materials. The most-productive and senior university scientists were the worst offenders, especially when they were involved in trying to bring their own research to market. And genetics researchers were significantly more likely than other life scientists to refuse faculty members access to their research results.

So widespread is the problem of secrecy among those working in the field of human genetics that an international group of molecular biologists in 1999 took the unprecedented step of pledging formally to release all DNA sequence data from all organisms into the public domain immediately. The pledge, which involved only researchers receiving public funds, came on the heels of a similar stand taken by the U.S. Human Genome Project, which is locked in a fierce competition with biotechnology companies to decipher the human genetic inheritance.

To ensure the free exchange of scientific information, many peer-reviewed journals that publish new research now insist that scientists make their technical materials and methods freely available to others as a condition of publication.

Even so, when neuroscientist Noel C. Harris at the University of Birmingham Medical School requested a sample of special cells featured in new research published in *Nature*, the scientists refused, in part because they were inhibited by their ties with pharmaceutical companies that might want to exploit the commercial possibilities of the experiment. Harris protested to *Nature* contending that the companies were trying to inhibit the ability of other scientists to work in the same field by restricting the distribution of research materials.[43]

"Secrecy about methods and results has become a common and accepted practice," said Steven Rosenberg at the National Cancer Institute.[44] "Just as a physician has a moral duty to do no harm, so does a scientist engaged in medical research. Deliberately withholding useful information or reagents is a violation of this principle. If secrecy slows research, then human suffering may be prolonged and unnecessary deaths may occur. Although these harms are not the intention of scientists who withhold information," Rosenberg said, "they are a logical consequence of such secrecy."

There is no shortage of sincere disagreements between scientists. When scientists make biased or misleading claims, however, they damage the effort to ensure that new medications or technologies are safe. They also undercut efforts to ensure that laws and public policies keep up with the cutting edge of science, disrupting efforts to assess controversial developments such as cloning and human embryo research.

The financial entanglements can also affect what is made public. When Tufts University researchers surveyed 1,000 scientists who wrote research papers in fourteen major scientific and medical journals in 1992, they discovered that a third of them had a direct financial stake in the research. Despite heightened awareness of such financial conflicts generally, many technical journals still decline to ask scientists about them or disclose them when publishing new research. Even the few journals that do require such disclosures rarely publish them.

Not until 1995 did the National Science Foundation and the U.S. Public Health Service require scientists to disclose any financial information bearing on a proposed research topic as part of an application for federal funding.[45]

A 1999 study by Sheldon Krimsky at Tufts and L. S. Rothenberg at UCLA examined the disclosure policies of the 1,400 most widely cited journals in 1997.[46] Of those, only 210 required contributors to disclose financial ties, such as stock ownership in the company whose drug was the subject of the research. Then, the researchers reviewed 62,000 articles published in those journals and found that in just 0.5 percent of the articles did authors acknowledge that they had some financial interest in the research topic.

Many scientists argue that, while perhaps worth noting for the record, such conflicts of interest are irrelevant because the facts of any experiment should speak for themselves. "Judge the science, not the author," editorialized Stephen J. Welch in *Chest*, the journal of the American College of Chest Physicians.[47] But several studies of biomedical controversies have shown that an expert's scientific position can be predicted by his financial relationships. Those who draw their support from the pharmaceutical or tobacco industry, for example, are more likely to draw conclusions favorable to their sponsors.

When scientists must struggle to balance research integrity and commercial advantage, it is often the public that suffers. Consider:

1. When University of Toronto researchers looked at the scientists who wrote journal articles supporting the use of controversial calcium channel blockers as a treatment for high blood pressure, they discovered that virtually all the scientists had a financial relationship

with the manufacturers of the medications.[48] The researchers looked at seventy articles published between March 1995 and September 1996—at the height of the controversy over the dangerous side effects of such drugs—and found that all the doctors who wrote in favor of the heart medications had received free trips, speaker's fees, or money for educational programs and research from the companies, compared to less than half of those who criticized the drugs. Only two of the seventy articles disclosed any conflicts of interest.

2. When Tufts University scientists who studied research looked at scientists who wrote pro and con about the controversial diet drug Redux, which was withdrawn for safety reasons in 1998, they found that researchers were far more likely to support the use of the drugs if they had a financial relationship with the drug's manufacturer than those who were neutral or critical of the obesity drugs.[49] When the *New England Journal of Medicine*, for example, published an editorial saying the benefits of such diet drugs outweighed the risks, it failed to note the authors had consulted for companies that made the drugs.

3. When the *Annals of Internal Medicine* published a research paper showing evidence that zinc lozenges helped relieve the symptoms of the common cold, the stock of Quigley Corp., which makes the tablets, jumped in response. Unbeknownst to the doctors and patients who read the study, the chief researcher held stock in the company. He sold it shortly after the findings were published with a profit of $145,000.[50]

4. When the *British Medical Journal* carried an editorial in April 1999 dismissing the idea that antidepressant drugs such as Prozac might be addictive, readers did not know that the company that makes the drug had given the authors free trips from Europe to Arizona in 1996 to attend a closed company symposium.[51] When other researchers checked 106 review articles of passive smoking, they found that the determining factor in whether the author found it harmful or not was whether or not the author was affiliated with the tobacco industry.

5. When the *New England Journal* published a series of medical reviews on new drug treatments for conditions such as multiple sclerosis, it did so without disclosing that in at least eight instances the authors had received research funds, travel money, or speaking

fees from the companies that made or marketed the drugs under review. The journal has perhaps the strictest disclosure standards of any peer-reviewed medical publication; even so the editor involved chose not to reveal the financial entanglements.[52]

As these issues continue to arise, the major medical journals have been redrafting their disclosure standards to maintain their credibility. The National Academy of Sciences, which offers scientific advice to the federal government on hundreds of technical and medical issues, also has changed its guidelines to promote greater openness about potential conflicts of interest that might bias its reports. Staff scientists, for instance, are required to report any conflicts they might have with industries affected by any of the academy's studies.

Nonetheless, the National Academy was still caught off guard when Michael Phillips, the staff scientist in charge of a crucial evaluation of the safety of genetically engineered crops, quit the academy in 1999 to go to work as a lobbyist for the trade association that represents some 800 biotechnology companies, including those that manufacture and market the genetically engineered crops.[53] Academy officials said that had they known of his intentions they never would have allowed the scientist to work on the study. Phillips's report, which was still being drafted when he resigned, was expected to have a major impact on how government regulators handle genetically engineered plants and produce.

Even the strictest disclosure standards cannot ensure that open debate over controversial research is open and unbiased. In early 1999, it was revealed that the Tobacco Institute had paid thirteen scientists a total of $156,000 to write letters challenging a major 1993 government report on secondhand smoke and lung cancer to prestigious research journals such as *Science*, the *Journal of the American Medical Association* (*JAMA*), the *Lancet*, and the *Journal of the National Cancer Institute* (*NCI*). The letter-writing campaign came to light in confidential legal documents released as part of Minnesota's $6.6 billion settlement with the tobacco industry. The documents cover a project called the ETS Consultant Program, which quietly recruited medical experts on behalf of the tobacco industry. "Ideal are people at or near retirement with no dependencies on grant-dispensing bureaucracies," one confidential memo noted.

For a letter to *JAMA* contesting an article on California workplace monitoring of secondhand smoke, the project allocated $5,000, the documents show. For a letter to the NCI journal disputing an editorial, the project paid $3,500. A letter contesting an editorial in the *Lancet* was worth $5,000. By concealing the industry's role in funding and editing the letters, the scientists essentially led the public to believe that the challenges arose from sincere scientific differences of opinion and not a hidden commercial agenda.

The scientists made no secret of their ties to the tobacco industry. But they did not disclose explicitly that they had been paid to write the letters, despite long-standing policies at the various publications that authors should disclose any potential conflict of interest. That, said Dr. Barnett Kramer, editor in chief of the NCI journal, "brings the issue to an entirely new level.[54] "It hadn't occurred to me that those sorts of things go on," Kramer said. "Now we ask."

The hidden price of stock options, consulting contracts, patent royalties, and commercial research grants may be that those scientists most familiar with the new science of life and the hazards it may pose will no longer feel free to dissent publicly. Certainly, the tension between contractual requirements of commercial secrecy and the need to inform the public about health hazards is more than some scientists can bear.

When, for example, Dr. Nancy Olivieri at the University of Toronto discovered a promising treatment for fatal iron deficiencies associated with hereditary blood diseases, she was elated.[55] Even before the first clinical trial, she published her findings in the *New England Journal of Medicine* in 1995.[56] Attracted by initial reports of success, a drug company called Apotex Inc. offered to help fund her research. She signed a nondisclosure agreement.

Then, almost as quickly, Olivieri and her colleagues developed doubts about the drug's toxic side effects. As disturbing data accumulated, she felt impelled to inform the children she was treating in two countries, update her patient consent forms, and then notify other researchers immediately. But the company sought to keep her concerns secret, invoking the nondisclosure agreement she had signed. When she presented her findings anyway, the university-affiliated hospital where she worked, which had been seeking a $20 million donation from the company, dismissed her as head of its blood disorders program.[57] Only after sustained pressure by faculty groups and leading health experts was she reinstated in 1999. The battle over the safety of the drug and whether it should be licensed continued unabated, but now the dispute was in the open. "I will never again accept drug company money that is not clearly and completely free of any secrecy clauses whatsoever," Olivieri said.[58]

When David Kern, an occupational health expert at Brown University, discovered evidence of a new deadly disease among workers at a nylon-flocking plant in Pawtucket, Rhode Island, he felt a duty to warn their employer.[59] The company, Microfibres Inc., hired him as a medical consultant. After he discovered eight more cases at two plants in two countries, Kern also felt obliged to alert other public health experts as well.[60] To his surprise, the company sought to suppress the study by arguing he was bound by a nondisclosure agreement he had signed on behalf of medical students taking an unrelated plant tour fifteen months before being hired as a consultant. At Memorial Hospital—a teaching hospital of the Brown Medical School—where he was chief of internal medicine, administrators agreed. A week after Kern spoke about the disease at an international medical conference, hospital officials announced they would not renew his contract and shut down his occupational health program. That also eliminated his place on the medical school faculty. "Various forces worked in an effort to prevent the dissemination of critically important scientific information that had immediate public health consequences," Kern said[61] "People were getting sick. It was not about curing a disease; it was about preventing one." At a recent medical meeting in San Diego, Kern reported five more cases of the disease.

When pharmacologist Betty J. Dong at the University of California, San Francisco discovered evidence that called into question the effectiveness of a thyroid medication taken daily by some 6 million Americans, she too found herself blocked from revealing her results by Boots Pharmaceuticals, the company that manufactured the drug and also paid for the $250,000 study.[62] Her research suggested the drug was no more effective than other, less expensive generic medications.

She, too, had signed a nondisclosure agreement. "It made me nervous," she said. "But evidently, a lot of people signed these clauses."[63] Not until 1997—seven years after her discovery—did her research reach the public through publication in the *Journal of the American Medical Association*. In the interim, the company had used her own data to publish a study that reached the opposite conclusions about the effectiveness of the drug, even as the firm insisted on her silence.[64] During that time, the company was purchased by Knoll Pharmaceuticals. Eventually, the company agreed to pay up to $135 million to settle a class action suit in San Francisco brought by patients who contended the company had overcharged them during the years it had suppressed Dong's research.

During a recent conference on secrecy in science, Robert Cooke-Deegan, director of the National Cancer Policy Board at the National Academy of Sciences, recalled a moment not so long ago when as part of the routine paperwork for a corporate research grant, he, too, was handed a nondisclosure form for his signature. It was a passport of sorts into the world of commercial biomedicine. "My whole life's work would have been wrapped up," he said. "The main criterion for ever releasing it would have been whether or not it harmed the company's business interests. And the company was the one to decide that." He decided not to sign.

For a well-established senior scientist, it is an easy decision. Self-interest and ethical considerations intersect. But for younger scientists who are beginning careers in an uncertain millennium such choices will certainly become more common, and researchers will find the commercial demands harder to ignore. Indeed, a number of prominent scientists have urged a renewed emphasis on a formal ethical code that should become an integral part of every scientist's ethos. In recognition that scientists can no longer in good conscience contend that their work is fundamentally amoral, Sir Julian Rotblat, the English physicist who won the 1995 Noble Peace Prize, recently urged the creation of a Hippocratic oath for scientists, in which researchers pledge to always consider the ethical implications of their work.

NOTES

1. Rex Dalton, "Charges Fly in $1bn Hormone Patent Battle," *Nature* 399 (27 May 1999): 289.

2. Ralph King, "Genentech to Remain Independent Firm—Roche will Buy Remaining 33 Percent Interest in Company, Then Sell Stock to Public," *Wall Street Journal*, 3 June 1999, A3.

3. "Academia vs. Industry: A US $1 Billion Hormone Patent Battle," *Science Week* 3, no. 24 (11 June 1999).

4. Paul Jacobs, "Stakes Are High in Genentech Patent Trial Over Growth Drug," *Los Angeles Times*, 3 May 1999, C1.

5. Associated Press, "UC Regents accept Genentech's $200 million offer," 19 November 1999.

6. W. Wayt Gibbons, "The Price of Silence," *Scientific American* (November 1996): 15.

7. S. M. Thomas et al. "Public sector Patents on Human DNA," *Nature* 388 (21 August 1997): 709.

8. FY 1998 Licensing Survey, Association of University Technology Managers Inc.

9. Steven Rosenberg, "Secrecy in Medical Research," *New England Journal of Medicine* 334, no. 6 (8 February 1996): 392–94.

10. David Blumenthal et al., "Relationships Between Academic Institutions and Industry in the Life Sciences," *New England Journal of Medicine* 334, no. 6 (8 February 1996): 368–73; David Blumenthal et al., "Withholding Research Results in Academic Life Science: Evidence from a National Survey of Faculty," *Journal of the American Medical Association* 227, no 15 (6 April 1997): 1224–28.

11. Constance Holden, "Postsecondary Education: Tenure Turmoil Sparks Reforms," *Science* 276 (4 April 1997): 24.

12. Rex Dalton, "Roche's Taq Patent 'Obtained by Deceit' Rules US Court," *Nature* 402 (16 December 1999): 709.

13. Eliot Marshall, "Trial Set to Focus on Peer Review: Cistron Biotechnology Sues Immunex Corp in Patent Case," *Science* 273 (30 August 1996).

14. "Immunex and Cistron Settle Patent Suit Over IL-1B Compounds," *Biotechnology Litigation Reporter*, 8 January 1997.

15. Interview with the "Hype and Profit: A Perilous Mix," *Los Angeles Times*, 24 January 1999, A1.

16. Ibid.

17. John H. Moore, "The Changing Face of University R&D Funding." *American Scientist* 86, no. 5 (September–October 1998): 402.

18. Robert Lee Hotz and Thomas H. Maugh II, "In Our Own Image: Life in a Genetically Engineered World. (First in a series)," *Los Angeles Times*, 27 April 1997, A1.

19. Ibid.

20. *Des Moines Register*, February 25, 1999

21. Hotz and Maugh, "In Our Own Image."

22. Dorothy Nelkin and Lori Andrews, "Homo Economicus: Commercialization of Body Tissue in the Age of Biotechnology," *The Hastings Center Report* (September 1998).

23. Hotz and Maugh, "In Our Own Image."

24. Blumenthal et al., "Relationships between Academic Institutions and Industry."

25. American Association of University Professors, press conference, Boston, Massachusetts, 17 May 1999.

26. "Report Paints Grim Outlook for Young Ph.Ds," Rand Corporation and Stanford University study, National Research Council report, September 1998. *Science* 281 (11 September 1998): 1584.

27. Constance Holden, "Report Paints Grim Outlook for Young Ph.D.s," National Research Council report, September 1998, *Science* 281 (11 September 1998): 1584.

28. Hotz and Maugh, "In Our Own Image."

29. Blumenthal et al., "Relationships between Academic Institutions and Industry."

30. "On the Eighth Day: A Series on Genetic Engineering: The New Breed: Scientists With An Eye on Profits," *Atlanta Journal-Constitution*, 1984.

31. Declan Butler, "Global R&D Spread Clouds Local Analyses," *Nature* 399 (6 May 1999): 6.

32. Robert Berring, "Is Berkeley Off Course," *California Monthly* 109, no. 4 (February 1999).

33. Irving A. Lerch, panel discussion, Secrecy in Science Conference AAAS/MIT, 29 March 1999. Robert Lee Hotz, "Secrecy is Often the Price Of Medical Research Funding," *Los Angeles Times*, 18 May 1999, A1.

34. John Deutch, panel discussion, Secrecy in Science Conference AAAS/MIT; 29 March 1999.

35. Interview with the author.

36. Marlene Cimons and Paul Jacobs, "Biotech Battlefield: Prohl vs. Public," *Los Angeles Times*, 21 February 1999.

37. "Owning Biological Discovery," *The Chronicle of Higher Education*, 5 February 1999.

38. Jon Cohen, "Report of New Hepatitis Virus Has Researchers Intrigued and Upset," *Science* 283 (30 July 1999): 644.

39. Holden, "Post-secondary Education," 24.

40. Holz, "Secrecy is Often the Price Of Medical Research Funding."

41. Richard Florida and Wesley Cohen, panel presentation, 1994 annual meeting of the American Association for the Advancement of Science.

42. David Blumenthal et al., "Withholding Research Results in Academic Life Science: Evidence from a National Survey of Faculty."

43. Noel C. Harris, *Nature* correspondence, 1999.

44. Steven A. "Sounding Board—Secrecy in Medical Research," *New England Journal of Medicine* 334, no. 6 (8 February 1996): 392–94.

45. *Tufts University Health & Nutrition Newsletter* May 1997.

46. S. Krimsky and L. S. Rothenberg, panel presentation, "The University/Industry Interface: Room for Improvements," at the 1999 annual meeting of the American Association for the Advancement of Science, 25 January 1999.

47. "Conflict of Interest and Financial Disclosure: Judge the Science Not the Author," *Chest* 112, no. 4 (October 1997).

48. Henry Stolfox et al., "Conflict of Interest in the Debate over Calcium-Channel Antagonists," *New England Journal of Medicine* 338, no. 2 (8 January 1998): 100.

49. *Journal of the American Medical Association.*

50. Krimsky interview.

51. Michael Day, "He Who Pays The Piper," *New Scientist* 9 May 1998, 18.

52. The Los Angeles Times, by Terence Monmaney

53. Melody Petersen, "Biotech Expert's New Job Casts a Shadow on Report," *New York Times,* 16 August 1999, A-10.

54. Interview with the author.

55. Interview with the author.

56. Nancy Olivieri, Gary Brittenham, Doreen Matsui et al., "Iron-Chelation Therapy with Oral Deferipronein in Patients with Thalassemia Major," *New England Journal of Medicine* 332, no. 14 (6 Arpil 1995): 918–22.

57. Richard Knox, "Science and Secrecy: A New Rift Proprietary Interests Found to Intrude on Research Disclosure," *Boston Globe,* 30 March 1999, A3.

58. Interview with the author.

59. Interview with the author.

60. David G. Kern et al, "Flock Worker's Lung: Chronic Interstitial Lung Disease in the Nylon Flocking Industry," *Annals of Internal Medicine* 129 (15 August 1998): 261–72.

61. Interview with the author.

62. JAMA. Interview with the author.

63. Interview with the author.

64. JAMA

ten

PROPRIETY AND PROPERTY
The Tissue Market Meets the Courts
LORI ANDREWS AND DOROTHY NELKIN

I n 1908 an Australian, Mr. Doodeward, acquired and preserved a two-headed fetus and exhibited it as a curiosity. The police confiscated the fetus, claiming Doodeward's actions violated public morality. When he sued to get the fetus back, the court was faced with a perplexing and far-reaching legal question that haunts jurists even today. Can someone claim to *own* a human body or its parts?

The need to devise an appropriate answer to that question is pressing, given the growing number of industries and relationships that utilize human tissue as the raw material for products. Biotechnology techniques have transformed a variety of human body tissues into marketable research materials and clinical products. Infant foreskin can be used to create new tissue for artificial skin. Eggs and sperm are bought and sold for both research and in vitro fertilization, and embryos have been stolen. Cell lines derived from the kidneys of deceased babies are used to manufacture a common clot-busting drug. Blood can serve as the basis for immortalized cell lines for biological studies and the development of pharmaceutical products; the catalogue from the American Tissue Culture Collection lists thousands of people's cell lines that are available for sale. Umbilical cord blood is a source of valuable stem cells for bone marrow transplantation. Blood has become one of the most valuable commodities on earth. While refined petroleum sells for $40 a barrel, an equivalent quantity of blood products is worth $67,000.[1]

New uses for human tissue are constantly being discovered. In November 1998 researchers funded by the biotech company Geron used excess embryos created through in vitro fertilization, which had been donated for research purposes, to create the most coveted of human cells: embryonic stem cells, primitive cells that can grow into every type of body tissue, including nerves, bones, and muscles. That same month, other Geron researchers reported that they had retrieved stem cells from the developing gonads of aborted embryos. In the future, biotech companies will use stem cells to produce marketable treatments—such as new heart cells to inject into an ailing heart, which would then repair itself.

And it's not just the medical sector that finds human tissue useful. Human bones, valued today as a means to study human history and satisfy curiosity, are stored in museums and sold in shops as biocollectables. Tissue such as blood, hair, and DNA—and even human corpses—are used as a medium by artists. Human DNA can also be used to run computers, since its ready replications provide more permutations than the binary code.

Due to the multiple uses in the biotechnology age, human body parts have increasing financial value. In fact, tissue has become so valuable scientifically, medically, functionally, and aesthetically that it has become subject to theft. Courts are struggling with questions of who should reap that economic gain and what constitutes an inappropriate taking of body tissue. Ironically, although the uses of human body parts have changed dramatically in the past two decades, courts' logic concerning ownership and control of the body has not changed very much since the obscure century-old Australian case.

OWNERSHIP AND CONTROL

The Australian court mediating the *Doodeward* dispute in 1908 was in a quandary. If the judge ruled the fetus was not Doodeward's property, the police might be able to confiscate bodies from archaeologists, collectors, and medical museums, among others. Unwilling to have that happen, the judge held, "It is idle to contend in these days that the possession of a mummy, or of a prepared skeleton, is necessarily unlawful; if it is, the many valuable collections of anatomical and pathological specimens or preparations formed and maintained by scientific bodies, were formed and are maintained in violation of the law."[2]

The court came up with an ingenious legal dodge, holding that if a corpse or a part of a corpse had been altered for use in medical or scientific examination, it acquired a value and became property. The Australian judge stated, "[I]f a person has by the lawful exercise of work or skill so dealt with such a body in his lawful possession that it has acquired some attributes differentiating it from a mere corpse awaiting burial, he acquires a right to retain possession of it, and, if deprived of its possession, may maintain an action for its recovery as against any person."[3]

Yet why should mere manipulation lead to a property interest? The *Doodeward* case itself should raise eyebrows. Nowhere in the opinion was the family of the fetus mentioned. Why didn't they have a claim or interest in the body? Why should Doodeward have property rights to a fetus when its own family members do not? The answer lies in the court's desire to protect certain social institutions.

Echoes of the *Doodeward* case resonate through the discussions and disputes about uses of body tissue today. Gunther von Hagens, a physician and anatomy lecturer at the University of Heidelberg, Germany, preserves and prepares corpses through an embalming process called "plastination." Plastination, which involves replacing the water in the tissue of a corpse with a polymer, provides the body tissues and organs with firmness and fullness so they can be displayed in a true-to-life manner. Von Hagens then arranges the corpses like sculptures. One startling figure is portrayed with all his organs and bones exposed, and his flayed skin draped over his arm like a robe. Another is stripped of his skin and flesh, leaving only bones and muscles. The dangling body of one man is dismembered, with the parts suspended on nylon strings. Another body is that of a five-month pregnant woman, revealing the fetus inside of her. There are bodies in familiar poses (running, sitting, gesturing), body parts in odd configurations, and pathological specimens (including abnormal human fetuses)—all in living color and on public display.

To obtain his bodies, Gunther von Hagens obtained consent from volunteers who knew how they would be used. Von Hagens did not accept the bodies of dead infants donated by their parents since the infants themselves could not give consent. However, he was able to obtain the bodies of infants who died thirty to forty years ago. He bought them from hospitals and medical schools that had preserved them for research and educational purposes.[4]

When asked who "owned" the bodies, von Hagens replied, "I do." He said that although there is no legal ownership of a corpse, his chemical transformation of a body (akin to that of Doodeward) makes it his property. "I have to insure it and pay taxes when it goes from border to border," he said. "I have to be able to claim it if it is stolen."

The question of ownership of body tissue routinely arises in the medical arena.[5] When John Moore, a Seattle businessman, fell ill with hairy cell leukemia, he went to a top specialist at the UCLA School of Medicine. He followed his doctor's orders, submitting to surgery to remove his spleen and other treatments. He returned to Seattle, thinking his disease was cured. But, for the next seven years, the UCLA doctor told him to keep flying back to Los Angeles for tests. Moore thought these visits were necessary to monitor his condition and complied out of fear that the leukemia might reappear. In reality, his physician was patenting certain unique chemicals in Moore's blood and setting up contracts with a Boston company, negotiating shares worth $3 million. Sandoz, the Swiss pharmaceutical company, paid a reported $15 million for the right to develop the cell line taken from Moore, which the doctors had named the Mo-cell line.

Moore began to suspect that his tissue was being used for purposes beyond his personal care when his UCLA doctor continued to take samples not only of blood, but of bone marrow, skin, and sperm. When Moore discovered that he had become patent number 4,438,032, he sued the doctors for malpractice and property theft.[6] Moore felt that his integrity was violated, his body exploited, and his tissue turned into a product: "My doctors are claiming that my humanity, my genetic essence, is their invention and their property. They view me as a mine from which to extract biological material. I was harvested."[7]

The case was so unusual that the trial court judge quickly threw it out of court. But Moore persevered and the California Court of Appeals ruled that he had been wronged. The appellate court underscored the growth of the market: "Until recently, the physical human body, as distinguished from the mental and spiritual, was believed to have little value, other than as a source of labor. In recent history, we have seen the human body assume astonishing aspects of value."[8] The appeals court reviewed cases involving celebrities like Bela Lugosi, who was held to have a property interest in his likeness, which prevented other people from marketing photos of him. "If the courts have found a sufficient proprietary interest in one's persona, how could one not have a right in one's own genetic material, something far more profoundly the essence of one's humanity than a name or a face?"[9] The court pointed out that the Uniform Anatomical Gift Act gives patients control over what is done with their bodies after they die, so it seems logical they should have control before they die. "Defendant's argument that the DNA from plaintiff's decision is not part of him over which he has the ultimate power of disposition during his life is . . . untenable," said the court of appeals.[10] On a more practical note, the court wrote, "If this science has become science for profit, then we fail to see any justification for excluding the patient from participation in those profits."[11]

The appellate decision was not the last word on the matter, though. The doctor and biotechnology company to whom he'd sold rights to Moore's cell line appealed to the California Supreme Court, where the deeply divided justices wrote eloquently about their individual views of the body, from sacred temple to biomedical factory. An acrimonious dispute surfaced in the four different opinions about whether or not Moore could claim that his tissue was his property. A majority of the justices, for differing reasons, rejected the idea that Moore could claim a property interest in his body. Even though the law in many ways was on his side, the majority seemed swayed by the heady promise of biotechnology. They didn't want to slow down research by universities or by the nation's biotechnology companies by "threaten[ing] with disabling civil liability innocent parties who are engaged in socially useful activities, such as researchers who have no reason to believe that their use of a particular cell sample is, or may be, against a donor's wishes."[12] They were concerned that giving Moore a property right to his tissue would "destroy the economic incentive to conduct important medical research."[13]

The research community also argued that Moore shouldn't have a property right in his tissue because he didn't have the ability to turn his bodily materials into socially useful products.[14] His doctor argued that he had transformed the cell line by immortalizing it—causing it to replicate indefinitely in culture. His claim to the tissue, like Doodeward's and von Hagens's, rested on the idea that he transformed it.

Although the California Supreme Court justices refused to recognize Moore's property right, they did toss him a legal crumb. They gave Moore a right to sue his doctor for breach of fiduciary duty and lack of informed consent. They held that a physician must tell his patient if he has a personal interest unrelated to the patient's health, whether research or economic, that might affect his judgment. "A physician who adds his own research interest to this balance may be tempted to order a scientifically useful procedure or test that offers marginal, or no, benefits to the patient," wrote Justice Edward Panelli for the majority.[15] This certainly seemed to be what happened to Moore, who was repeatedly called back to his doctor's office to give blood, bone marrow, sperm, and other tissue.

Using the fiduciary duty approach only protects patients from *physicians* who collect the tissue and not other researchers or employers and other institutions that may be collecting or testing patients' tissue samples without their knowledge or consent. And, in an even larger loophole, it does not provide protection if the patient was sick—as Moore was—and so probably would have undergone the operation anyway, no matter what the doctors were planning to do with his tissue.

"Reversing the words of the old song," wrote Justice Stanley Mosk in dissent, "the nondisclosure cause of action thus accentuates the negative and eliminates the positive: the patient can say no, but he cannot say yes and expect to share in the proceeds of his contribution."[16]

John Moore is now one of a growing number of individuals whose cell lines you can order by perusing the American Tissue Culture Collection catalogue, or its foreign counterparts.[17] The entries describe, in telegraphic style, the people whose tissue is for sale: CRL 5867, a forty-nine-year-old black female with cancer of the lymph node; JCR B0068, a fourteen-week-old Japanese fetus who died of cytomegalovirus; SK-HEP-1, a fifty-two-year-old German man with liver cancer. Thousands of individuals are listed, but it is doubtful that more than a handful of them or their families realize that they are part of this elite market.[18]

MODERN BODY-SNATCHING

The body business is becoming a pivotal, though often unnoticed, part of our economy. In the genetics sector alone, there are about fifty private DNA testing centers in the United States, hundreds of university laboratories undertaking DNA research, and over one thousand biotechnology companies developing commercial products from bodily materials.[19] Large, publicly traded pharmaceuticals companies depend on human tissue. But even smaller, one-product companies are listed on the stock exchange. When John Buster, then an infertility specialist at UCLA Harbor General Hospital, joined with venture capitalists to sell human embryos to potential parents, they listed their company on the NASDAQ exchange under the moniker "BABY."[20]

Many of the products made from bodily material may ultimately yield useful treatments for patients, but there are serious problems in the way they are developed, distributed, and turned into commodities. When commercial interests and the quest for profits are a driving force, questions of human safety and respect for the human sources of tissue—the person in the body—take second place.

The modern situation is reminiscent of body-snatching in the nineteenth century. At that time, because bodies were in short supply, they became valuable commodities. Anatomy departments paid between ten and thirty-five dollars for a body, more than the weekly wage of a skilled worker at that time.[21] Bodies were obtained in devious ways—through grave robbing and even the murder of beggars.[22] As described by historian Ruth Richardson, corpses were "quarried": "Parts extracted were sold to those who could use them, such as dentists and wigmakers, and to those who assisted medical research and study, such as articulators of bones for medical skeletons, and medical-specimen makers. Profits were to be made at every stage."[23]

The potential use of organs and tissue for transplant and research has escalated traffic in body parts. Not only have infertility doctors been accused of appropriating reproductive tissue, but scores of coroners, morgue workers, and physicians have removed organs and other tissue without patients' consent to sell for transplantation. Body parts have even been stolen from the site of accidents and sold to meet the demands of industry and medicine.[24] In a sting operation, the Food and Drug Administration (FDA) nabbed some "body brokers" in the business of importing human tissue. Posing as representatives of a tissue bank, FDA agents ordered tissue from a California dentist who tried to sell the body parts at a discount.[25] In France, a government investigation exposed an embezzlement scheme in which private companies billed local hospitals for synthetic ligament tissue that, it turned out, came from human tissue; in France, human tissue legally cannot be bought and sold.[26]

Funeral home personnel and medical examiners in recent years have engaged in tissue theft. In one case, a morgue employee allegedly stole body parts and sold them nationally—a situation uncovered unexpectedly when the body of a twenty-one-day-old infant was exhumed for other purposes and found to be missing his heart, lungs, eyes, pituitary gland, aorta, kidneys, spleen, and key brain parts.[27] Mortuaries and crematories in California routinely, and without consent, sold body parts and organs from deceased individuals to the Carolina Biological Supply Company.[28] The company turned a blind eye to how the tissue had been collected.

Organs and tissues, in short supply, have been raided elsewhere as well. In France in 1992, Christopher Tesniers was killed in an accident, and his eyes were removed without consent. His parents, billed for postmortem surgery, found that other body parts had also been removed.[29] When they expressed their dismay to the media, the number of French families willing to donate loved ones' organs plummeted.

Even government agencies, in the United States and abroad, have taken tissue in inappropriate ways. The British Department of Health and Medical Research Council paid mortuary attendants for the pituitary glands from 900,000 corpses to extract human growth hormone.[30] The British Human Tissue Act only allows such removal if there is no objection from the family.[31] The council argued that the act did not require them to ask consent, but only to avoid taking the tissue if the family had specifically objected.[32] Yet, as philosopher Baroness Mary Warnock noted, "There might be religious persons who would be outraged that their child has had their pituitary gland removed."

The takings came to light due to side effects of the treatment made from human growth hormone extracted from one of the pituitary glands. In the 1960s and 1970s, nearly 2,000 congenitally short British children were injected with human growth hormone from the corpses to make them taller and "liberate [them] from a life of taunts and humiliation."[33] By 1996, though, eighteen of the children had died of Creutzfeld-Jakob disease (CJD), a progressive brain disorder which can take up to twenty years to manifest.[34] The tissue had transmitted the disease. The hormone was withdrawn from the market without the parents being told why.[35] Ultimately the families won compensation by showing that the council knew of the dangers and kept using the hormone.

The black market in tissue has multiple impacts. If not prevented, theft of tissue can create emotional, social, and financial problems for the tissue source. Personal and social views of the body serve certain functions for individuals and their communities. A person's control over what is done to her body, or its parts, is important to the individual's psychological development and well-being. It is also a means of establishing identity and conveying values to others. But body tissue has social importance beyond the individual. Social conceptions of the body establish community identification, encourage socially responsible behaviors, and set acceptable priorities for group activities.

People obviously care about any physical intrusions on their bodies, and the criminal and civil laws of battery serve to protect them. But there is also evidence that people care about what is done to their tissue even once it is no longer part of their bodies. In light of past research exploitation, some African American women refuse to allow amniotic tissue to be collected for prenatal diagnosis out of concern about the uses that could be made of this tissue.[36] Patients in France who provided tissue for research to a nonprofit institute later protested against the proposal that the samples be sold to a for-profit company.[37] Other patients have objected to their cells being patented.[38]

Even the prospect of research on a person's tissue after death has a potential psychological impact on the person while he is alive or on the next of kin. Living individuals' psychological interests are the foundation for numerous legal rules that give control to people over what is done with their property or corpses after they die. Some commentators may argue that people can no longer be harmed once they are dead, but inheritance laws and restrictions on the use of organs after death are based on the psychological impact on people while alive.[39] Polls indicate that one major reason why people do not donate organs is that they do not like the idea of being cut up after their death;[40] another is that many people want to be buried intact.[41] Yet laws in many states authorize research on corpses without previous consent.[42]

Unauthorized taking of tissue can also violate religious beliefs. Jewish tradition, for example, maintains that as man was created in the image of God, in death the body should retain the unity of that image.[43] In the Orthodox Jewish community, the body must be buried whole.[44] If a person's leg is amputated during his or her life, facilities exist to store that body part for burial with the individual after death.[45] Taking an individual's tissue without consent would violate that belief.

Taking tissue without consent for commercial purposes can be viewed as a form of theft. Yet the question of how to handle those who appropriate body parts has vexed prosecutors for centuries. The problem is that the body has not been considered the "property" of the person under the law, so it has been hard to punish as theft. Nevertheless, in each generation, as the body and its parts take on new uses and new values, the need for an appropriate legal doctrine to cover body theft has grown.

The law in most states has not caught up with these new developments. Since the body has not traditionally been considered property, biocrime is not a legally recognized offense. Instead, prosecutors and civil litigants either have to ignore the appropriation of tissue or find some other way to address it, such as through claims of intentional infliction of emotional distress, public health laws, a specific tissue protection law, or laws about outraging public decency. All of these other approaches, though, have their limitations for dealing with abuses that occur involving research on and commercialization of body tissue.

EMOTIONAL DISTRESS

People have an interest in what will happen to their extracorporeal body parts while they are alive and even after they die. Yet, in civil litigation, protection of that interest now tenuously rests on precarious doctrines that protect people from emotional distress. In a 1973 case, a physician at Columbia Presbyterian Hospital in New York City attempted to fertilize a woman's egg in vitro with her husband's sperm. Without consulting the physician or the couple, the department chairman removed the culture from the incubator and destroyed it. The couple, Mr. and Mrs. Del Zio, sued the department chairman and the hospital's trustees, charging conversion of personal property and intentional infliction of emotional distress.[46] The jury rejected the property claim but awarded the plaintiffs damages for the emotional distress. Mrs. Del Zio got $50,000 for emotional distress and Mr. Del Zio, $3.00. It is ironic that Mr. Del Zio, who provided his sperm to the mixture, was compensated so poorly and not considered to have a property interest. He could have sold that sperm to an infertility clinic for much more than $3.00.

Even emotional distress damages are often not recoverable. Courts in many states allow people to receive compensation for emotional harm only if it accompanies an actual physical injury—for example, if the person is in a car accident.[47] But theft of embryos or other tissue outside of the body doesn't physically injure the patient. In addition, in many states, there is a cap on the amount recoverable for emotional distress, and the individual or couple may not be adequately compensated.

Relatives' emotions are also underprotected by emotional distress laws.

When a deceased person's remains are disturbed in a grave, a relative can sue for harm, in the words of one court, because of "the tender sentiments uniformly found in the hearts of men, the natural desire that there be repose and reverence for the dead, and the sanctity of the sepulcher."[48] But recovery can only be granted if the "conduct was so extreme and outrageous as to go 'beyond all possible bounds of decency' and was . . . 'utterly intolerable in a civilized community.'"[49] There may be a lot of leeway to use a corpse's tissue—such as in making a commercial medical treatment—contrary to the individual's prior wishes but, nonetheless, for a goal that is not "utterly intolerable."

THE PUBLIC HEALTH APPROACH

Public health laws, too, are inadequate to prevent misuses of body tissue. On September 16, 1997, federal and New York State enforcement officials executed a search warrant on William Stevens's store, Evolution, in Soho. They found dozens of items made from endangered species remains, including a tiger skin rug, ashtrays made from gorilla feet, a gibbon arm, a bald eagle skull, and a stool made from an elephant foot. Federal officials charged Stevens with a felony for trafficking in endangered species; he faced a potential eleven-year imprisonment and $600,000 in criminal fines.[50]

The police also found rows of human fetuses in jars. Yet Stevens did not face prosecution for selling them. Rather, he was charged with a misdemeanor for transporting human remains—the fetuses—into New York without a public health permit.[51] The only reason officials targeted Stevens was that a jar containing a fetus broke in transit and the United Parcel Service (UPS) had noted something leaking through the package.

Similarly, food and drug regulations have been used as a basis to arrest individuals who were importing human placenta from China, misleadingly labeling it as a "medical herb" to be sold here in stores and clinics.[52] Such public health laws do not provide sufficient protection for people who are the sources of the tissue. A woman whose placenta or fetus is taken in violation of her religious beliefs and subsequently sold would have no recourse. Moreover, courts seem to feel that anything doctors do with human tissue is okay under the public health laws. Moore's doctor's commercialization of Moore's cell line was not an approved manner of dealing with human tissue under the California public health law, yet the court supported his patenting the cell line to further the growth of biotechnology.[53]

The public health law at issue in the *Moore* case allowed "scientific use" of body tissue. Dissenting Justice Stanley Mosk pointed out that "It would stretch the English language beyond recognition, however, to say that *commercial* exploitation of the kind and degree alleged here is also a usual and ordinary meaning of the phrase 'scientific use.'"[54] The majority in *Moore*, though, dismissed this argument saying "philosophical issues about 'scientists bec[oming] entrepreneurs' are best debated in another forum."[55]

SPECIFIC LAWS

At the University of California at Irvine, doctors were accused of taking eggs from infertility patients and, without the patients' knowledge or consent, implanting them in other women. In the wake of this scandal, the California legislature enacted a tissue-specific bill that makes it a crime to steal human eggs.[56] But this law is hardly cause for celebration. First, it seems ludicrous that such a law would be necessary; why wasn't the theft of reproductive materials already covered by existing criminal laws? Second, it provides no protection for *other* body parts like blood, genes, or cell lines that might be subject to misappropriation.

Nor do statutes about outraging public decency protect the sources of body tissue or affect health care professionals' use of tissue. They have mainly been applied to people interfering with the burial of corpses—or displaying dead bodies publicly.

When artist Richard Gibson decided to create a sculpture that "was intended to show the place of humans in society and how we treat human beings," he took an unusual tack.[57] He obtained two human fetuses, each of three- to four-months' gestation, from an anatomy professor.[58] He then sculpted a head with the fetuses hanging as earrings. Gibson and the owner of the British gallery displaying his work were prosecuted for the common law offense of outraging public decency.[59] The defense lawyer asked that the men be charged under the Obscene Publications Act of 1959 because it allows a defense for art that is in the public good.[60] The prosecution refused the request, and the men were convicted by a jury. The judge said, "I accept that your motives were genuine. But in a civilized society there has to be a restraint on the freedom to act in a way that has an adverse effect on other members of society."[61] The court fined Gibson $875 and the gallery owner $612.[62]

The main difficulty with a prosecution for outraging public decency is that it only applies to a few situations. Like the legal cause of action for emotional distress (which is about offending a particular individual, rather than the public), it requires proof that the use of the tissue was outrageous—beyond that permissible in a civilized community. Medical exploitation and commercialization of a patient's tissue without consent is likely to be found justifiable by the courts, as it was in the *Moore* case. The anatomy professor who gave Gibson the fetus, for example, was not prosecuted. Moreover, the "outrageous conduct" charge protects the public at large but fails to grant rights to the individuals whose tissue was used. The women whose fetuses became earrings in the *Gibson* case were not even mentioned in the legal opinion.

A MOVE TOWARD THE PROPERTY MODEL

Because existing laws have serious deficiencies, a few courts in recent years have begun to address whether body parts should be considered the property of the individual or, if the individual has died, his next of kin. Courts' comfort with using a property approach has evolved slowly over the past decade. The courts have been troubled about the implications of what an unbridled property approach might mean. At the same time, they have increasingly realized the need for greater protection of people whose body parts have gained in value in the biotech era.

Part of the concern about giving people a property right in their tissue is the concern that *other* people might treat it as property. If that happens, a man applying for welfare might be turned down because his kidney is worth $20,000. A college student may be denied a student loan because her eggs could be sold for anywhere from $2,500 to $50,000. A hospital may try to take, sell, and use eggs from a comatose woman to cover her hospitalization costs. Considering such dark possibilities, courts and other policymakers should make clear that, even if an individual can treat her body as property, others should not be able to do so.[63]

Such concerns are legitimate, but could be avoided by a property approach that allows the individual alone to determine what should be done to his body tissue. Limitations on what can be done to property are common in other legal situations. As Susan Rose-Ackerman points out, many forms of property have restrictions on alienability.[64] There may be restrictions on who holds them, what actions are required or forbidden, and what kinds of transfers are permitted. Some types of properties can be given as gifts, but not sold (items made of the fur or feathers of endangered species, for example). Other types of properties (such as the holdings of a person who is bankrupt) can be sold, but not given as gifts.

The advantage of the property approach is that it closes the gap in protections. Paul Matthews, pointing out that dead bodies and body parts are not considered goods under English law, asks, "If the ashes of X, a celebrity, are without consent 'removed' and (say) later auctioned at a London auction-house, can anything be done by X's next of kin, or personal representatives?"[65] Matthews argues that bodies and body parts should be characterized as property so that interference with them would be prohibited.

Some courts are gingerly moving toward the property model. Risa and Steven York froze an embryo in a Virginia clinic on their fourth attempt at in vitro fertilization and then moved to California. They wanted to transfer the embryo to their new doctor there, a physician who had achieved pregnancies for other couples shipping their embryos to his clinic.

They explored how to safely move the embryo across the country and learned that it could be transported—like other human tissue or organs—in a liquid nitrogen tank. Steven, a physician, had transported human tissue for transplants and so felt completely comfortable about picking up the embryo and transporting it back to California. Risa called the Virginia clinic to make arrangements to pick up her embryo. The physician who took the call told her he would not allow it. Risa and Steven were stunned. "They're holding my baby hostage," Risa said.

The Yorks sued the doctor, citing the emotional distress they felt at losing what might be their best chance to become parents. Risa was approaching forty, and the eggs she could produce now might not be as likely to develop into a child as the embryo in the Virginia clinic. The court, though, responded with the general legal rule: unless the doctors' actions *also* caused physical harm to Risa, she couldn't recover for her emotional devastation.

The Yorks would have been out of luck if the judge had not been willing to apply property law. The court said the same legal principle that allows people to get their car back from a parking garage gave the Yorks a right to their embryo. This 1989 decision may have only a limited application beyond that case, though, since the clinic's informed consent form had actually referred to the embryo as the couple's "property."

A year after the Virginia court decided the *York* case, an Ohio court struggled with another body case, *Brotherton* v. *Cleveland*.[66] After pronouncing her husband dead, hospital personnel asked his surviving spouse, Deborah Brotherton, if she wanted to donate his organs. She refused based on her husband's aversion to organ donations. However, in spite of her refusal, the coroner permitted the deceased's corneas to be removed for transplantation. The hospital documented her refusal in its "Report of Death" but did not directly communicate her refusal to the coroner. The coroner didn't bother to review the hospital's report. The court ruled in favor of Deborah Brotherton. In deciding the case, the court did not go so far as to say that the corneas were property. But it did conclude that "the aggregate of rights granted by the state of Ohio to Deborah Brotherton rises to the level of 'a legitimate claim of entitlement' in Steven Brotherton's body, including his corneas, protected by the due process clause of the fourteenth amendment."[67]

By 1995, five years after the *John Moore* case, a court finally admitted that tissue outside a person's body could be considered property. Anthony Herrera was an assistant to the pathologist who conducted autopsies at the Saginaw, Michigan, Community Hospital. Herrera's job was to open up the bodies and then sew them back up after the pathologist had finished his work.

Yet Herrera had another job—he owned and operated the Central Michigan Eye Bank and Tissue Center. So when the autopsies were over, without the consent of the next of kin, Herrera would pop out the deceased's eyes. Then, he would sell them out of his eye bank.[68] When relatives of his deceased victims later sued, the trial court dismissed their claims. However, the appeals court was clearly troubled that people might not be protected if bodies could not be considered property in some instances. If a woman's husband died in the neighbor's yard, one justice asked, should the neighbor simply be able to keep the body?[69]

The answer to the appellate court was clear. In its ruling, the unanimous court said emphatically, for the first time, that the next of kin have "a constitutionally protected property interest in the dead body of a relative."[70]

THE EVOLUTION OF BODY LAW

Every generation of courts has been forced to address the legal categoriza-
tion of bodies and body tissue as other social institutions make claims to
them. The logic of the *Doodeward* case—saying that professionals who transform bodies can own them—reverberates through the courts even today.
But judges are becoming increasingly uncomfortable with saying that
people have no property interest in their own bodies and tissue since they
may have legitimate reasons to want to direct the use of their physical parts.
Courts are realizing that a person who has a kidney removed to donate to
a sibling would be wronged if the doctor gave the kidney to someone else.
They are recognizing that, if the person's body tissue is tested genetically
without consent for identification, paternity, or disease detection, the person
may suffer psychologically, socially, and financially as a result.

A major sign of judicial discomfort with the law's reluctance to recognize property interests in the body appeared in a 1998 British case.
Though the case concerned the theft of body parts for art, the court was
especially troubled that the current law did not protect against the types of
abuses that were likely to arise as transplantation and genetic testing
become more prominent.

In that case, artist Anthony Noel-Kelly convinced Richard Heald of
the Royal College of Surgeons to grant him access to the institution so he
could sketch cadavers. Such access was in keeping with a long-standing
artistic tradition; in fact, Heald introduced Kelly to his colleagues as a
modern-day Leonardo da Vinci.[71]

Soon, though, sketching and painting watercolors were not enough.
Kelly wanted to make casts of the body parts, which would require taking
them to his studio. Since that was beyond the scope of his agreement with
Heald, Kelly hatched a plan to steal them, enlisting the aid of Neil Lindsay,
a technician at the college.[72] Lindsay would pull the body parts out of the
vats of formaldehyde, remove the identification labels, wrap the parts in bin
liners, and transport them in a rucksack by taxi or the Underground to
Kelly's flat. The thefts lasted from June 1991 to November 1994. Kelly's
loot included three heads, three torsos, part of a brain, six arms or parts of
them, and ten legs and feet.

Once in his studio, Kelly used rubber molds, fiberglass, and plaster to make a copy of the body parts and then buried the body parts in the grounds around his family's estate.[73] One of Kelly's statues for sale was a chunk of a woman's body dissected to reveal part of her womb.[74] Another was the head and torso of an old man with a hole drilled in his head; a £4,500 price tag was put on that work.[75]

Unfortunately for Kelly and Lindsay, Kelly's gallery exhibit of his death images also gained the attention of a police officer who realized that the pieces had been cast from dead bodies, and they were charged with theft. They defended themselves with an argument—one also used by doctors and scientists who do research on human tissue—that the body parts had been "abandoned" in the college's basement storeroom.[76] They also argued that they were treating the corpses better than doctors or the royal college had. "Each piece I treated with respect," said Kelly. "I did not cut them up with a saw."[77] He argued that the individuals whose bodies he used would have approved of his actions. "I felt if the donors were looking on, I was not insulting their bodies in any way."[78]

Lindsay, too, argued that the royal college itself was not treating the body parts appropriately since they had been in the possession of the college in some cases for more than twenty years longer than a new law allowed, and they should have been buried. Kelly and Lindsay tried to portray themselves as rescuers—the "route to a proper burial" of human remains that were not properly being handled by the medical community.[79] Their main legal argument, though, was a deceptively simple one of wide application. They argued that they could not legitimately be charged with theft because, for centuries, judges had said the human body could not be property.

In considering the *Kelly* case, the British judges pondered whether anyone could "own" a dead body or its parts. For centuries, the legal answer had been no. That's why the turn-of-the-century body snatchers who stole corpses were charged not with theft, but with outraging public decency.[80]

The prosecutor in Kelly's case, Andrew Campbell-Tiech, said the "no property in a body" rule had been the result of an erroneous reading of a 1614 case in which the defendant had disinterred corpses to steal their burial clothes. "Generations of lawyers," he said, then perpetuated the error.

Even though the legal doctrine had a "very poor legal pedigree," noted Judge Geoffrey Rivlin, he could not just reverse centuries of legal tradition. Judge Rivlin admitted it was strange that body-snatching in its heyday had not been declared a felony.[81] It was treated lightly in the nineteenth century, even though there were more than two hundred other offenses garnering the death penalty, ranging from shooting rabbits to appearing in disguise on a public road.[82]

"Why was body-snatching not made a felony?" asked Judge Rivlin in his opinion. "It is difficult to resist the thought that in not bringing the full weight of the law down upon the practice, Parliament, if not exactly turning a blind eye to it . . . winked at it in the interest of medical science." And policymakers are still winking at the interests of medical science today. The *Moore* case, for example, allowed the doctor, but not the patient, to assert a right in the patient's tissue.

Kelly and Lindsay tried to use this legal loophole to their advantage, by claiming they couldn't be charged with theft of property when body parts were not, after all, property. The judge agreed that body parts were not property under the common law, but cited the 1908 *Doodeward* case for the principle that altering body parts—as the royal college had done by putting them in formaldehyde—turned them into the royal college's property. Kelly and Lindsay were convicted on April 3, 1998. They were the first British citizens to actually be prosecuted and convicted for theft of human remains.[83]

Like the *Doodeward* case itself, though, any mention of the family members of the deceased was totally absent, even though there was plenty of evidence that the royal college was acting improperly itself by keeping body parts without burial longer than the anatomy act permitted.[84]

Ironically, as part of the litigation, Kelly's molds and sculptures were turned over to the Royal College of Surgeons, which could display them in its museum.[85] This underscored the fact that medical institutions are allowed to do things with body parts that other individuals and institutions are not. But this unquestioned privileging of the medical sector may be as outrageous to many people as were Kelly's acts.

In fact, it is ironic to think that while individuals generally cannot have a property interest in their bodies, scientists are quick to claim a property interest in patients' cell lines. Such a claim, after all, was the basis of a six-year dispute between microbiologist Leonard Hayflick and the National Institutes of Health (NIH) over ownership of a cell line Hayflick had developed from embryonic tissue using NIH funds.[86] In a similar case, Hideaki Hagiwara, a postdoctoral biology student at the University of California, San Diego, urged faculty member Ivor Royston to fuse cancer cells from Hagiwara's mother with another cell line to create a hybridoma that would produce human monoclonal antibodies.[87] Once the combined cell line was created, Hagiwara smuggled some cells out of the lab and into Japan to treat his dying mother. But he also applied for a patent in Japan on the hybridoma, which he intended to distribute through a company owned by his father (the Hagiwara Institute of Health). Hagiwara felt his family had an economic interest in the new cell line since he proposed the project and his mother provided the original cells. Royston disagreed, arguing that he and his colleagues invented the procedure and created the parent cell line that made the production of the human monoclonal antibodies possible. In part, Royston was concerned that Hybritech, a biotechnology company in which he was a shareholder and which funded some of his work on hybridomas, would lose out on its claim to an exclusive license if Hagiwara could assert an interest. Hagiwara and Royston ultimately reached an agreement, giving the University of California the patent and the Hagiwaras an exclusive license for the cell line in Asia.[88]

In thinking ahead about the ways in which body tissue might be used in the biotech age, the British judges in the 1998 *Kelly* case envisioned that, as we further enter the biotechnology age, courts might treat even unaltered body parts as property if they have a significance beyond their mere existence. "This may be so," wrote the court, "if, for example, they are intended for use in an organ transplant operation, for the extraction of DNA, or for that matter as an exhibit in a trial."

If that new logic were followed, sufficient new protections would be added for the use of tissue samples for research, information, and commercial purposes. Giving people a property interest in their body tissue would allow them to have greater control over how that tissue is used, before and after their deaths.

NOTES

1. Richard Bernstein, "A Science and a Business, a Saver and a Killer," a review of *Blood: An Epoch History of Medicine and Commerce*, by Douglas Starr, *New York Times*, 23 September 1998, E7.

2. *Doodeward v. Spence*, 6 CLR 406, 413–14 (1908).

3. Ibid. The case was about a two-headed fetus.

4. Edmund L. Andrews, "Anatomy on Display and It's All Too Human," *New York Times*, 7 January 1998, A1.

5. See Lori Andrews and Dorothy Nelkin, *Body Bazaar: The Market for Human Tissue in the Biotechnology Age* (New York: Crown, 2001).

6. *Moore v. Regents of the University of California*, 793 P2d 479 (Cal 1990).

7. John Vidal and John Carvel, "Lambs to the Gene Market," *Guardian* (London), 12 November 1994, 25.

8. *Moore v. Regents of the University of California*, 249 Cal Rptr 494, 505 (1988).

9. Ibid., 508.

10. Ibid., 507–508.

11. Ibid., 249 Cal Rptr 494, 509.

12. *Moore v. Regents of the University of California*, 793 P2d 479, 493, 51 Cal 3d 120, 143 (1990).

13. *Moore v. Regents of the University of California*, 793 P2d 479, 495, 51 Cal 3d 120, 146 (Cal 1990).

14. It should be noted that scientists should have no higher claim under this argument since, as individuals, they themselves couldn't transform the tissue. They need millions of dollars in equipment, generally subsidized by taxpayers in the form of NIH grants. Even their individual expertise is socially subsidized. Taxpayers, not the students, pay the bulk of costs of medical training.

15. *Moore v. Regents of the University of California*, 793 P2d 479, 484, 51 Cal 3d 120, 130 (1990).

16. Ibid., 181.

17. http://www.atcc.org; for a foreign database, see, for example, http://wdcm.nig.ac.jp/DOC/menu3.html.

18. John Moore's cells are for sale as CRL-8066; a plasmid containing Moore's DNA sequence that codes for colony stimulating factor is sold as ATCC 39754.

19. According to the biotechnology trade organization BIO, there are 1,283 biotechnology companies in the United States. "Some Facts About Biotechnology," <http://www.bio.org/whatis/editor_welcome.html>.

20. Lori B. Andrews, *The Clone Age: Adventures in the New World of Reproductive Technology* (New York: Holt, 1999), 37.

21. Michael Sappol, "The Cultural Politics of Anatomy in 19th Century America: Death, Dissection, and Embodied Social Identity," (Ph.D. diss., Columbia University, 1997), 528.

22. Ibid.; see Ruth Richardson, "Fearful Symmetry, Corpses for Anatomy: Organs for Transplantation," in *Organ Transplantation: Meanings and Realities*, ed. Stuart J. Youngner et al. (Madison: University of Wisconsin Press, 1996), 66, 82.

23. Richardson, "Fearful Symmetry, Corpses for Anatomy."

24. "The Body-Parts Trade," *World Press Review* 41, no. 4 (April 1994).

25. *CBS Evening News*, 21 May 1996.

26. Catherine Tastemain, "Oversight for Tissue Transplants," *Nature-Medicine* 1, no. 5 (May 1995): 397.

27. Frank J. Murray, "Survivors May Sue Over Theft of Body Parts," *Washington Times*, 6 November 1995, A1.

28. The Supreme Court of California recognized an action for emotional distress. The court also pointed out that the supply company could be liable as well, since it "knew or should have known that desecration of human remains would necessarily occur." *Christensen* v. *Superior Court of Los Angeles County*, 54 Cal 3d 868, 878 (1991).

29. Tara Patel, "France's Troubled Transplant Trade," *New Scientist*, 3 July 1993, 12–13.

30. See Elizabeth Grice, "The Lost Children Who Longed to be Like Everyone Else," *Daily Telegraph*, 29 March 1996, 27.

31. Dominic Kennedy, "Families Challenge Legality of Trade in Pituitary Glands," *Times* (London), 4 September 1995, 6.

32. See Kennedy, supra note 31. The statute itself is ambiguous. It says, "The person lawfully in possession of the body of a deceased person may authorise the removal of any part from the body for use for the said purposes if, having made such reasonable enquiry as may be practicable, he has no reason to believe—that the deceased had expressed an objection to his body being so dealt with after his death, and had hot withdrawn it; or that the surviving spouse or any surviving relative of the deceased objects to the body being so dealt with." Human Tissue Act 1961 (c. 54), Section 1, Removal of Parts of Bodies for Medical Purposes.

33. Elizabeth Grice, "The Lost Children Who Longed to Be Like Everyone Else." Parents who sued the Department of Health and Medical Research Council argued that if they had known of the risk, they would have avoided the treatment and let their children remain short. They also said that the Department of Health continued using the treatments for years after its dangers were known. The High Court ultimately ruled in favor of the patients' families on the grounds that treatment given after July 1, 1985, was negligent since by then the Medical Research Council knew of the risks. The hormone was extracted from the pituitary glands of corpses; it took one hundred glands to create enough hormones for a year's treatment for a single child.

34. "Milking the System?" *Lawyer*, 19 November 1996. Symptoms include blindness, deafness, memory loss, and death within a year. Anna Pukas, *Timebomb Killer*, 2 September 1993, 7.

35. See Grice, supra note 32.

36. Rayna Rapp, "Refusing Pre-natal Diagnosis: The Uneven Meanings of Bioscience in a Multicultural World," *Science, Technology and Human Values* 23 (1998): 45–70.

37. Declan Butler, "French Geneticists Split Over Terms of Commercial Use of DNA Bank," *Nature* 368 (1994): 175.

38. Philip R. Reilly, Mark F. Boshar, and Steven H. Holtzman, "Ethical Issues in Genetic Research: Disclosure and Informed Consent," *Nature Genetics* 15 (1997): 16–20.

39. Obviously, there are certain religions in which the individual *can* be harmed—such as being condemned to hell if his dead body is treated improperly.

40. "For the Record," *Washington Post*, 13 January 1987, A22. See also Christopher Heredia, "The Ultimate Offering," *San Francisco Chronicle*, 11 January 1999, A3.

41. "Survey: Most Americans Willing to Donate Organs," *United Press International*, 30 March 1993, Regional News. See also Carmen M. Radecki and James Jaccard, "Psychological Aspects of Organ Donation: A Critical Review of Synthesis of Individual and Next-of-Kin Donation Decisions," *Health Psychology* 16 (1997): 183–95.

42. Fourteen states have such laws. See Chapter 10, "Regulating the Body Business."

43. Maurice Lamm, *The Jewish Way in Death and Mourning* (New York: Jonathan David Publishers, 1969), 10.

44. Ibid.

45. Henry Fitzgerald, Jr., "Woman Awarded $1.25 million in Suit: Funeral Home Must Compensate for Losing Mother's Amputated Legs," *Sun-Sentinel* (Fort Lauderdale), 16 May 1997, 1B. In fact, when Menorah Gardens and Funeral Chapels lost an amputated leg of an Orthodox Jewish woman, it made a $1.25 million lawsuit settlement to her daughter. "Orthodox Jews believe that at the end of time, not only will a person's soul be resurrected, but the body as well. . . . It's important that the whole body, including blood, be buried."

46. *Del Zio v. Manhattan's Columbia Presbyterian Medical Center*, No. 74-35558 (SDNY 14 November 1978).

47. William L. Prosser and W. Page Keeton, *The Law of Torts*, 5th ed. (St Paul, Minn.: West Publishing, 1984), Chap. 9, 364-65. See also 38 *American Lawyer Reports* (1985) § 1003 et seq.

48. *Carney v. Knollwood Cemetery Assn.*, 514 NE2d 430, 432 (Ohio 1986).

49. Ibid. at 433.

50. See *United States v. William Stevens*, F # 1998 R00O38 (complaint filed by U.S. Attorney Ilene Jaroslaw in the Eastern District of New York).

51. See *People v. Stevens*, Supreme Court of the State of New York, Supreme Court Information, March 1998.

52. See "J&J International Systems Hit," 4 *Washington Information Source - Warning Letter Bulletin*, 17 June 1996 (No. 12, ISSN: 1069-4218). There is no information reported about what, if any, penalties the FDA imposed on Chi Thanh Trading. Nor did it state how the company came to obtain the human placentas, or from whom they came.

53. *Moore v. Regents of the University of California*, 51 Cal. 3d 120, 793 P.2d 479, 271 Cal. Rptr. 146, 161 (1990).

54. 271 Ibid., 175 (Mosk, J., dissenting).

55. Ibid., 159 n.34.

56. California Penal Code §§367f,g went into effect 1 January 1997.

57. "Two Face Charges in Britain for Showing Human Foetus Earrings," *Reuter Library Report*, 31 January 1989.

58. Gibson claims that an anatomy professor gave him the fetuses and told him that they were more than twenty-five years old. "Foetus Earrings Made to Promote Debate Says Artist," *Daily Telegraph*, 8 February 1989, 3.

59. The offense carries a maximum penalty of life imprisonment. See "Two Face Charges in Britain for Showing Human Foetus Earrings," supra note 57.

60. Section 4 of the Obscene Publications Act of 1959 provides that "a person shall not be convicted of an offense against section two of this Act . . . if it is proved that publication of the article in question is justified as being for the public good on the ground that it is in the interests of science, literature, art or learning, or of other objects of general concern."

61. John Weeks, "Art Pair Fined over Foetus Earrings," *Daily Telegraph*, 10 February 1989, 3.

62. "British Jury Finds Fetus-Skull Earrings an Outrage," *Reuters*, 9 February 1989. Gibson appealed on the grounds that section 2(4) of the Obscene Publications Act of 1959 precluded a prosecution for outraging public decency. That section provides: "A person publishing an article shall not be proceeded against for an offense of common law consisting of the publication of any matter contained or embodied in the article where it is of the essence of the offense that the matter is obscene." The Court of appeals upheld the conviction. *Regina v. Gibson and Another*, 1 All ER 439 (Court of Appeal, Criminal Division, 1990). It ruled that because "obscene" under the Obscene Publications Act (Section 1 of the Act provides: "For the purposes of this Act an article shall be deemed to be obscene if its effect . . . is, if taken as a whole, such as to tend to deprave and corrupt persons who are likely, having regard to all relevant circumstances, to . . . see . . . the matter contained or embodied in it. . . .") means corruption of public morals, it only stops prosecution under the common law for actions that corrupt public morals, but does not stop every offense that falls under an outrage to public decency charge. Thus, Gibson

could be convicted under the common law offense of outraging public decency; he did not have to be tried under the Obscene Publications Act. The court of appeals also held that the prosecution did not have to prove intent to outrage public morals or reckless, it just had to prove that the intentional publication took place. Furthermore, the artist or gallery owner did not have to try to draw people in to look at the exhibit, it was sufficient for conviction that the mannequin was displayed.

63. Nor should judges be allowed to sentence offenders to pay their fines in body product donations (once the property approach has established a market value for them). If this seems farfetched, consider that there already have been instances in which judges sentenced defendants to give blood transfusions. Similarly, an eighteenth-century British statute allowed judges to order anatomical dissection of hanged murderers. Paul Matthews, "Whose Body? People as Property," in Lord Lloyd of Hamstead and Roger W. Rideout with Jacqueline Dyson, *Current Legal Problems 1983* (London: Steven & Sons, 1983), 205. In 1890 a man sold the Royal Caroline Institute in Sweden the rights to his body after death. Later, he tried to refund the money and cancel the contract. In the subsequent lawsuit, the court held that he must turn his body over to the institute and also ordered him to pay damages for diminishing the worth of his body by having two teeth removed. Russell Scott, *The Body as Property* (New York: Viking, 1981), 185–86. In contrast, in the United States, under the Uniform Anatomical Gift Act, promises to donate body parts upon death are revocable. With living donors, revocation should be allowed up until the time the transfer is made.

64. Susan Rose-Ackerman, "Inalienability and the Theory of Property Rights," *Columbia Law Review* 85 (1985): 931.

65. Matthews, "Whose Body? People as Property," 207.

66. 923 F2d 477 (6th Cir 1991).

67. Ibid., 478.

68. *Whaley v. County of Tuscola*, 58 F3d 1111, 1112 (6th Cir 1995).

69. *Whaley*, 1116.

70. *Whaley*, 1116.

71. "A Leonardo Preoccupied with Death," *Birmingham Post*, 4 April 1998, 2.

72. The judge gave Lindsay less jail time because he was younger than twenty-one when the thefts began. The judge also recognized Kelly's strong personality and influence over Lindsay.

73. Kathryn Knight, "Sculptor Jailed for Theft of Body Parts," *Times* (London), 4 April 1998, 5.

74. "Jury Sees Body Sculptures," *Guardian*, 27 March 1998, 7.

75. Kelly said that the sculptures were a way of "demystifying death." See "A Leonardo Preoccupied with Death," supra note 71. "I find beauty in death . . . these

rotting bodies," said Kelly. "You look at them and remind yourself, this is how we all end up." John Windsor, "Death is No Sleeping Beauty," *Independent*, 5 April 1998, 5. His work struck a responsive chord in people, gaining the admiration, for example, of Prince Charles. Tony Ellin, "Burke and Heir," *Daily Record*, 4 April 1998, 4.

76. Melvyn Howe, "Body Parts Taken on London Tube to Artist's Studio," *Press Associated Newsfile*, 31 March 1998.

77. Paul Cheston, "I've No Qualms about Going into a Morgue, I Find Beauty in Death," *The Evening Standard*, (London) 3 April 1998.

78. "Dismembered Corpse on the London Tube," *Belfast News Letter*, 1 April 1998, 7. Kelly claims that he always had respect for the body parts; however, in his diary, he referred to one severed leg stored in his attic as "Hopalong" and he did not bury the parts in any type of container. Melvyn Howe, "Judge Jails Sculptor for 'Revolting' Thefts," *Belfast News Letter*, 4 April 1998 3 [hereinafter Howe, "Judge Jails Sculptor"].

79. Howe, "Body Parts Taken on London Tube to Artist's Studio"; *Regina* v. *Kelly and Another*, 3 All ER 741 (Court of Appeals, Criminal Division, 1998).

80. Melvyn Howe, "How One Can Steal What Cannot Be Owned," *Press Association Newsfile*, 3 April 1998.

81. Ibid.

82. Ibid.

83. "Body Parts Sculptor is Freed From Jail on Appeal," *Evening Standard* (London), 15 May 1998, 21. Previous body snatchers had been charged with a lesser offense—outraging public decency. Melvyn Howe, "Judge Jails Sculptor"; Kelly and Lindsay appealed, arguing that the trial judge erred in ruling that body parts were property. The Court of Appeals, Criminal Division, upheld the conviction: "[I]n our judgment parts of a corpse are capable of being property within section four of the Theft Act, if they have acquired different attributes by virtue of the application of skill, such as dissection or preservation techniques for teaching or display purposes." *Regina* v. *Kelly and Another*, 3 All ER 741 (Court of Appeal, Criminal Division 1998). However, the appellate court reduced Kelly's nine-month sentence to three months. "Body Parts Sculptor Is Freed from Jail on Appeal," *Evening Standard* (London), 15 May 1998, 21. Kelly was released on May 15, 1998, after he had served half of the sentence. See ibid. Lindsay's sentence was reduced to two months, rather than six months, and was suspended for two years. The justices considered the facts that both defendants were supposedly not motivated by malice or greed (although Lindsay was paid for his help in stealing the bodies, and the sculptures that Kelly made were offered for sale at ten times the price that Kelly paid Lindsay), that they had never been in trouble before, and that Kelly was motivated by "artistic reasons." Cathy Gordon and Jan Colley, "Bodys-

natching Artist has Sentence Reduced," *Press Association Newsfile*, 14 May 1998, Home News section. However, the justices were concerned that the theft could have a "dissuading effect" on someone who was thinking about donating his body for research. See ibid. In addition they thought that the many people would view the acts "with repugnance." See ibid. "We accept that however questionable the historical origins of the principle," said the court, "it has now been the common law for 150 years at least that neither a corpse nor parts of a corpse are in themselves, and without more, capable of being property protected by rights. . . . If that principle is now to be changed it must be by Parliament." *Regina* v. *Kelly and Another*, 3 ALL ER 741 (Court of Appeal, Criminal Division). See also Gordon and Colley, "Bodysnatching Artist Has Sentence Reduced."

84. The lawyers for Kelly and Lindsay argued that, even if taking the body parts might be considered theft in some instances, it couldn't be a crime here because the Royal College of Surgeons was not the owner of the parts. The Anatomy Act of 1832 made it legal for anatomists and certan others to possess body parts and corpses. Among other things, the anatomy act requires a person or institution to have a license to legally possess a body. Kelly and Lindsay said since the college was violating the anatomy act by keeping body specimens longer than its license permitted, it did not legally possess the body parts, and the defendants were not stealing the parts because they did not belong to someone else. Section 5(1) of the Theft Act says: "Property shall be regarded as belonging to any person having possession or control of it, or having in it any proprietary right or interest. . . ." Legally, samples could only be retained for three years and then had to be disposed of, yet some of the samples were over twenty years old, and were not being used by the college. The court held that the royal college could not prove it was in lawful possession of the corpses, it only had to show that it had possession when the body parts were taken for the defendants to be guilty of theft.

85. Melvyn Howe, "Artist Jailed for 'Revolting' Theft of Human Parts," *Press Association Newsfile*, 3 April 1998.

86. Constance Holden, "Hayflick Case Settled," *Science* 215 (1982): 271.

87. Marjorie Sun, "Scientists Settle Cell Line Dispute," *Science* 220 (1983): 393; see also "A Flap Over Patient Profit Sharing," *National Journal* 18 (1986): 1530.

88. Lori B. Andrews, "My Body, My Property," *Hastings Center Report* 16 (October 1986): 28–38.

eleven

PROPERTY RIGHTS AND HUMAN BODIES
PILAR N. OSSORIO

N ew discoveries in medical science and biotechnology have propelled us into a historical moment in which the uses of human-derived bodily materials are multiplying rapidly. More people have interests in these materials than ever before. Human-derived bodily materials are increasingly seen as scarce resources—we are told that there is a shortage of transplantable organs,[1] that there is a shortage of oocytes (eggs) for studies in human development and for infertility treatments, and that there is a shortage of fetal cells available for experiments, such as those to treat or cure Parkinson's disease.[2] This unsatisfied demand has led some theorists to argue that we should now allot property rights in human-derived bodily materials and create or expand markets through which these materials, and rights to them, can be exchanged. Markets, these theorists argue, will lead to an economically efficient distribution of human-derived bodily materials; they will end up in the hands of those who value them most.

However, in Anglo-American law and philosophy, and in public discourse, we have been reluctant to conceive of humans or their body parts as property. We look back on those contexts in which we did so, such as the slavery of black people in the United States, with shame and sorrow. That we no longer consider children to be their parents' property or wives to be their husbands' property is perceived as social and moral advancement. And although next of kin have certain rights with respect to the disposition of deceased people, the law rarely labels these rights as property rights. Many arguments against applying property concepts to human-derived bodily materials derive from the claim that it is inhumane and a violation of human rights for one human being to own another.

In this essay I will first examine what we mean by the words "property" and "ownership," then I will evaluate some arguments that are commonly made about who owns human-derived materials. Finally, I will examine the relationship between the concepts of "commodification" and "property," because I think the most threatening aspect of property in bodies or body parts is the idea that it will likely lead, via some form of slippery slope, to commodification of live, intact human persons.[3]

Throughout this essay I use the terms "human-derived bodily materials" and "bodily materials" quite broadly to encompass human DNA, cells (including reproductive cells such as eggs and sperm), tissues, and organs. I also include fetuses, embryos, and corpses. I will indicate instances in which claims refer to a specific kind of bodily material, or exclude some types.

PROPERTY AND OWNERSHIP

"Property" and "ownership" may at first seem like self-evident and uncomplicated concepts. I bought my car with good cash money and I own that. I bought my house. Well, I'm buying it, and the bank still owns a far greater share than I do. But between the bank and me, it is our property. I own those old photos I inherited from my grandmother—the ones of my father when he was a teenager with his hair cut short and his ears sticking out. But, do I own my education? Do I own my reputation? Do I own an entitlement to welfare benefits? Do I own the right to pollute? Do I own the right to vote? Do I own myself? Do I own my grandmother's dead body? Do I own cancer cells that have been removed from my body? Do I own frozen embryos if they were conceived using my gametes? These last questions are ones on which scholars, courts, policymakers, and ordinary people disagree.

Although notions of "property" are deeply ingrained in our society (perhaps in all human societies), there is no unitary definition of the term; there is no definition that applies in all instances. Property has been described as reflecting a triadic relationship among institutions, resources, and culture.[4] It has also been described as a form of power over others by virtue of the capacity to dictate the use of resources.[5] At the very least, we can say that property concerns relationships among people with respect to things or ideas.

Scholars do generally agree that property is not an object, property is not the thing itself, but rather a bundle of rights in an object. This may seem a bit counterintuitive—when you own a car, your property is not the car itself, but rather the rights you have in the car. Rights generally included in the property bundle are the right to possess or occupy; the right to use; the right to exclude others from possessing, using, or occupying; and the right to transfer (to sell or give away—to alienate).

Rights in the property bundle are rarely, if ever, absolute. Nor is it essential for all to be present for something to be denominated as property. You may own a car, and yet your use rights are constrained by speed limits, drunk-driving laws, and licensing requirements such as the requirement that you be a certain age before you drive. You may own land, yet your right to destroy vegetation growing on that land will depend on environmental laws and whether that vegetation is a common dandelion or an endangered species. So, one important point is that even if we call something property, that by no means ends the discussion about what may be done with the thing.

Also, rights typically present in the property bundle can be present even in the absence of a property right. We each have rights to use public side-walks and public buses, but we do not own them. I can occupy and use public beaches, but sadly, that does not mean I own any beachfront property.

There is also no clear definition for the terms "own" or "ownership," in part because different bundles of rights are associated with owning different kinds of property. Ownership is not reducible to any singular set of rules. The rights I have when I own a car are different than the rights I have when I own a pet, and different again than rights in land. It is clear, however, that sociocultural assumptions about ownership set the background against which legal rules operate.

The concept of property as a bundle of rights is important to help us articulate and respond to concerns about rights in bodies and human-derived bodily materials. For instance, some people who object to the concept of property in human cells, tissues, or DNA are concerned about the moral permissibility of a vast range of experimentation on humans or human tissues. Such people would like to see use rights curtailed or heavily regulated. They may fear that if we assign property rights to bodily materials, then the default assumption will be that owners of such materials have robust and extensive use rights. Then, those who wish to regulate might have a heavier burden of persuasion in arguing for regulation.

The right to transfer that is typically part of the property bundle also concerns many people when applied to human bodily materials. Some are concerned that an unfettered right to transfer materials such as reproductive cells and embryos by selling them in fairly unregulated markets will lead to an inappropriate application of marketplace values to human beings (discussed in more detail below).

Patents are a distinct form of property that may apply to human-derived bodily materials such as DNA and cell lines. Patents differ from most other forms of property in that (1) the rights conveyed by them are entirely enumerated in federal statutes; and (2) these rights are very limited when compared to most other property rights. Patents convey only the right to exclude others from making, using, selling, or importing the patented item, or the right to exclude others from carrying out a patented process. Patents do not convey positive rights to possess, make, use, or sell anything! I could have a patent on a radar detector, and yet it could be illegal for me to possess, make, use, or sell one. I could have a patent on a genetically modified form of corn, and yet it could be illegal for me to grow or market that corn. Because patents do not convey rights to possess or use, it can be misleading to refer to patent rights as a form of ownership—ownership generally connotes possession and use rights.[6]

Patent rights also have time limitations; they begin when the patent issues and run twenty years from the day the patent was filed.

WHO OWNS ME?

Some who oppose the application of property rights or property rhetoric to human bodily material offer as a reason the claim that because it is wrong for one person to own another, it is therefore wrong for one person to own another's body parts. On the other hand, some commentators believe it self-evident that a person owns any materials derived from her body. A reason offered to justify this claim is that I own myself, and therefore I own all of my parts and anything derived from my body. Commentators who take this position object not to the possibility that human bodily materials may be considered property, but that our legal regimes and cultural norms generally do not assign initial ownership rights in extracorporeal body parts to the person from whom those parts (DNA, cells, tissues) were derived.

There are several questions to be addressed in attempting to sort out who, if anybody, should have property rights in human-derived bodily materials. These include: Do I own myself? If I own myself, does it mean that I own my body? If I do not own myself, does it mean that I do not own my body? If I own myself, does it follow that I own my constituent parts? If I do not own myself, does it follow that I do not own my constituent parts? If I own my constituent parts, does it follow that I own those parts when they are not attached to my body?

With respect to the question "Do I own myself?" we should first note that the premise (a) "nobody else owns me" does not logically compel us to accept the conclusion that (b) "therefore, I own myself." Conclusion (b) would only follow from premise (a) if the world were such that I must be owned by somebody. However, even if we accept that another person cannot own me, it is also possible that *nobody* at all owns me; I do not even own myself. If somebody denies that I own myself, he is not necessarily implying that I am a slave; perhaps I and other persons are not the sort of entity to which property concepts reasonably apply.[7]

How things become property is a puzzle. Once property rights attach to an item, it is not difficult to understand how property is transferred from one person to another—by sale, gift, or devise. But "every chain of title must have a first link," and the difficult question concerns the manner in which we initially come to apply property concepts to objects.[8] Among philosophers and other scholars, there is great disagreement about whether there are facts about the world from which it follows as a matter of logic, justice, or other ethical imperative that an individual ought to be accorded property rights in anything. And it is certainly a matter of dispute whether or not persons are the sort of entity about which it can be said, "I own myself."

The philosopher John Locke stated that "Every man has a property in his own person. This nobody has any right to, but himself."[9] On the other hand, philosophers Immanuel Kant and Georg Wilhelm Friedrich Hegel believed that it makes no sense to speak of owning oneself because a person is a subject in whom the ownership of other things can be vested. Ownable things are objects, not subjects, and therefore the concept of owning oneself is contradictory. As Kant put it: "But a person cannot be property and so cannot be a thing which can be owned, for it is impossible to be a person and a thing, the proprietor and the property."[10]

But perhaps we need not answer the question "Do I own myself?" to answer the question, "Do I own my body?" Certainly, the relation of the self to the body is not clear from the preceding quotes. One could believe that "nothing is more one's own than one's body"[11] while still believing that a person cannot own herself in the sense that she cannot own her personal identity. Views about the relationship of self to body vary tremendously. Some believe that the person, the "I," is distinct from the body. Call this view I_1. Whereas others believe that the person is just constituted of, or can be reduced to, biological material and the conscious experiences made possible by that biological material. Call this view I_2. There are numerous intermediate positions between I_1 and I_2. For instance, one could believe that persons are constituted of biological material and a spirit.[12]

Those who hold I_1 or some similar view can argue that a person's body is her property without running into Kant's conundrum of collapsing the subject and the object, the proprietor and the property. Under I_1 the I who owns the body is distinct from the body; there is some me for whom my body can be an object, a possession. Those who hold I_2 must find a way to overcome Kant's conundrum if they are to justify the claim that "I own my body."

A great deal of our collective reluctance about applying property rules and property rhetoric to humans or our parts is that it threatens to remove humans from the realm of agents and subjects. To think of ourselves or our parts as property attaches a label that typically signifies things that are passively acted upon, things used for the ends of others rather than entities that have ends of their own. We are deeply suspicious that we cannot hold in our consciousness conceptions of humans or our parts as property and still respect each other's or our own humanity and human rights.

Having recognized that philosophers are deeply divided about whether we can attach property concepts to ourselves, we can also ask how the law conceives of our rights in our bodies and body parts. How do my legal rights with respect to my body compare to a typical property rights bundle? By law we recognize people's bodies as nearly inviolate; so long as a person has not been arrested, there are very few justifications for doing something to a person's body against his will, or compelling him to submit to bodily invasions. We have constitutional rights to be secure against unreasonable searches and seizures, not to be deprived of life or liberty without due process, and not to be enslaved. Courts generally will not force a person to submit to a medical procedure even to save another person's life—"[F]or our law to compel defendant to submit to an intrusion of his body would change every concept and principle upon which our society is founded."[13] Although, some lower courts have ordered pregnant women to undergo forced caesarean sections.[14]

In addition, if a person signs a contract in which she agrees to personally perform a service, such as repairing a roof, but she fails to perform the service and is sued for breach of contract, courts generally will not order her to perform the service. Rather, courts will order the defendant to pay enough money to put the aggrieved party in the position he would have been in had the defendant not breached. Courts do not order people to complete personal services contracts because that would be too similar to involuntary servitude or slavery.

From this cursory review of legal rights, we note that individuals have legal rights to exclude others from using or possessing our bodies. The right to exclude others from using or possessing is typically part of the property bundle. However, when it comes to our bodies, exclusionary rights are often conceived of as rights of personal privacy and autonomy rather than property rights.

Of course, one of the important rights in the property bundle is the right to transfer. Do we have rights to transfer ourselves by gift or sale? Is it even coherent to ask this question? There are respects in which we are legally barred from selling ourselves or otherwise creating a market in human beings—we cannot sell ourselves or others into slavery, and in theory we cannot sell sexual services or babies. But we can sell our intellectual and physical labor. And if you follow professional sports, you will recognize that there are some respects in which there are markets for people or at least for their talents, markets in which athletes are bid upon, exchanged, and valued in dollars—lots of dollars. All of this is done to maximize the preference satisfaction of team owners. Whether or not markets for athletes or their talents are degrading and dehumanizing is a matter of public debate.

Another arena in which courts frequently confront the question of whether there can be an explicit property right in human bodies concerns rights with respect to human corpses. The courts have found that next of kin have certain rights of disposal that include the right to determine whether the body is buried, cremated, or preserved in a legally approved manner. Absent some prior arrangement by the deceased (such as a prepaid burial plot), the next of kin can determine the place of burial and change it at their pleasure. At least in one instance, a court held that if a place of burial is taken by the state for public use, the state indemnifies the next of kin for the costs of moving and reinterring the remains.

Although these rights involve some use, possession, and exclusion, nearly all U.S. courts have held that they do not "rise to the level" of property rights. They are sometimes referred to as quasi property rights. As the Georgia Court of Appeals put it when discussing the mystery of human death and the disposition of corpses: "While it may be true that our laws must govern the funeral procession of a corpse, it will not impose a pecuniary value on the flesh itself. To do so would make the strangest thing on earth that much stranger."[15]

In short, people have rights in their bodies that look a bit like property rights, but we are reluctant to apply property concepts directly and openly to human bodies, and for some, the concept of even self-ownership is incoherent. An area in which there is near universal agreement, however, is that it is morally wrong for one person to own or attempt to own another. If people or their bodies are ownable, they are ownable only by themselves and not by other people.

WHO OWNS MY BODY PARTS?

If I own myself, or if nobody owns me, then who owns my constitutive parts? Does it follow from the fact that I (a person or an intact corpus) may not be owned by others that my extracorporeal body parts may not be owned by others? Does it follow from the fact that I may not be owned by others that my extracorporeal body parts should be owned first and foremost by me? Most of the pressing policy questions have to do with whether, or under what conditions, scientists, doctors, and biotechnology companies have rights to possess, use, or sell extracorporeal human bodily materials. Should those who want to make scientific use of human bodily material be required to buy that material from research subjects or patients, or at least negotiate an explicit transfer of rights? Do research subjects and patients abandon or discard their bodily materials when these materials are removed from their body, and do we therefore lose any right to negotiate about them once they become extracorporeal? Are extracorporeal bodily materials *res nullius*—owned by none and available for appropriation? Then perhaps it is the case that scientists make those extracorporeal bodily materials into property by using human ingenuity and labor to turn the materials into something different than they naturally were and something more economically useful.

One type of argument that is put forth in favor of the proposition that research subjects and patients should be said to own materials taken from their bodies can be stated as follows: (1) my living body may not be owned by others; (2) my cells are part of my living body; (3) even when my cells are not attached to my body, they are part of me; and therefore (4) my cells may not be owned by others. For this argument to be convincing, we would need to make sense of proposition (3). In what way are my cells "part of me" when they are no longer attached to my body? Are they part of me in a way that the reasons for prohibiting ownership of people's bodies make sense when applied to extracorporeal cells?

Very briefly, we may not own other people or their bodies because people have their own life plans, their own goals, their own purposes, their own "sakes." People have wills. For one person to have robust property rights in the being or corpus of another, for one person to own another, would be to deny or suppress the owned person's will by allowing the owner to inject her own life plans, goals, and purposes into the owned person. The owned person's will would be excluded or disavowed by that of the owner. The owned person would be turned into an object. For reasons that should be obvious, nearly all societies of the world have agreed that treating a person as an object is dehumanizing and a violation of human rights.

These reasons justifying prohibitions on one person owning another apply to live, intact bodies. They do not so clearly apply to extracorporeal bodily material. Cells do not have a will of their own, organs do not have a sake, and although some people would disagree, I contend that fetuses, embryos, and cadavers do not have life plans, purposes, or goals of their own. Not yet, or not anymore. My intact corpus constitutes me to a greater or lesser degree (depending on one's view of personhood). If somebody could control my body, they could frustrate and undermine my personhood. It is not so clear that the same could be said about another person owning my extracorporeal bodily materials. True, most bodily material that derives from me will contain my DNA (although my reproductive cells or an embryo/fetus made from them would only have half of my DNA), and they will contain proteins, lipids, and other chemicals that were once within my body. But to hold that our personhood is contained in any particular bit of extracorporeal bodily material would be to take a very reductionist view of personhood, and perhaps to invest more meaning into DNA, cells, or organs than is healthy or reasonable.

If somebody is doing experiments with my cells (possessing and using them), this will not prevent me from attending college, marrying the person of my dreams, or starting a business. If a researcher creates an immortalized cell line from my cells and the licenses that cell line to a biotechnology company for a million dollars, that will not prevent me from traveling on my vacation, writing a play, or getting a pet. It is quite unlikely that there is a global sense in which the mere fact of another person possessing, using, transferring, or excluding others (even me) from my extracorporeal bodily material will annihilate my will, my goals, or my life plans. It simply does not follow that because my intact corpus or my personal identity should not be owned by another, that my extracorporeal bodily material necessarily should not be owned by another.

There are some ways in which another person's ownership of my bodily materials might impede or undermine specific goals I hold. For instance, suppose I believe that research on the genetics of intelligence should not be done, but somebody used DNA isolated from my white blood cells to conduct such experiments. Or suppose I believe that human DNA should not be patented, but after isolating an interesting gene from my cell, a scientist created artificial copies of this gene and then patented his creation. In these scenarios material from my body is being used to thwart my goals, or to advance somebody else's goals that directly conflict with my own. Even if I do not hold extremely reductionist views of per- sonhood, it might still be possible for me to conceive of the use of my bodily materials in these contexts as some way in which I was forced to contribute to a project that I opposed.

A second example in which another person's property rights in my extracorporeal bodily materials could undermine my goals, projects, or life plans is if stigmatizing or otherwise negative information about me was derived from those cells and linked back to me. For instance, suppose a sci- entist determined that my DNA contained a mutation making it highly likely that I would develop early onset Alzheimer's disease. If this infor- mation became accessible to employers or insurers, it could damage my opportunities for employment, and it could diminish my insurability.

Thus, while it seems misguided to hold that another person's ownership of my extracorporeal bodily material is like slavery, it is still the case that another person's ownership of those materials could be detrimental to me in various ways. These problems could arise even in the absence of acknowledged property rights in human bodily materials, if research were done without my consent, or if proper privacy and confidentiality protections were not maintained. And, even if third parties had property rights in my bodily materials, those rights could be regulated and restricted in a manner that would diminish the likelihood of these detrimental impacts.

There are views about personhood, and about the relationship of personal identity to external objects or to body parts, that would support arguments against another person's owning my extracorporeal body parts. For instance, there is a view of personhood in which almost any kind of object external to one's body can become constitutive of one's identity by virtue of its importance in one's life or life plans. This view is held by Hegel and his intellectual followers. For instance, a wedding ring, given as a symbol of two people's union and worn as a symbol and reminder, can become constitutive of its wearer in a manner that takes it out of the pure object world and includes it in the person's identity.[16] Such a wedding ring would have a value to the wearer that goes far beyond the amount paid for it, or the amount she could obtain were she to put it on the market. She would not willingly exchange her ring for another one that was exactly the same or of the same monetary value. To her, the particular wedding ring has become "priceless."

In that situation the wedding ring is no longer "mere property." It has become part of her identity, and for anther person to lay property claims to it while it was part of her identity could be dehumanizing. Objects that are incorporated into our identities are objects that we do not exchange in the market.

I can envision circumstances in which extracorporeal body parts could take on a similar priceless character. Suppose that I have my kidney removed so that I can donate it to my brother, or my bone marrow removed so that it can save somebody else's life. In these instances, the extracorporeal bodily material might have significance for me that goes beyond its material significance or any monetary value it might have. My sense of who I am may become bound up with that material, and its significance may actually surpass that which it had when it was a mere internal constituent of my body. Under such circumstances, the possibility that somebody might possess and use those extracorporeal bodily materials for purposes other than those I had intended, or that physicians might choose to give my kidney to somebody other than my brother to advance their own goals, could undermine my goals and my will so severely as to compromise my identity and my overall psychic well-being.

We should recognize, however, that under this theory personal identity is quite fluid.[17] Objects can be "me" at times and "not me" at other times. This would apply to bodily materials as well as anything else. Thus, under such a theory, it is also plausible that once a kidney or bone marrow had been removed from my body, I would have little sense of connection to it. My bodily material could become just another object, deserving of no more respect than a rock or a car on the used car lot. At most, theories of the fluid boundaries of personal identity cannot tell us that certain external objects, including extracorporeal bodily materials, are always sacrosanct or always to be treated as unimportant with respect to personal identity. The theory can tell us that there may be times in which it can sensibly be said that a person's extracorporeal bodily material is still constitutive of his identity in a strong sense, in a sense that makes it problematic for another to claim property rights in the bodily materials.

Another instance in which the possibility of another owning my extracorporeal bodily material would be problematic is if I adhered to the belief that my soul or spirit infused every portion of my body. If I believed this, then my extracorporeal materials would still be "me" in some fundamental, nonreductionist way. To have somebody else possess, use, or sell them would not necessarily frustrate my life plans or undermine my will, but it would, perhaps, damage my spirit and my opportunities for an afterlife.

Thus far I have argued that from the proposition "nobody else may own me," it does not follow that "nobody else may own my extracorporeal bodily material." The usual reasons justifying prohibitions on one person owning another do not apply to one person owning bodily materials derived from another. And although there are views of personal identity under which extracorporeal bodily material might plausibly be seen as constitutive of a person's identity in a manner that precludes ownership of that material by another, these views and the reasons supporting them would not apply to most people most of the time. But, if it is not clearly wrong for another person to own my extracorporeal bodily material, is it wrong for this person to claim ownership without a clear transfer of rights from me? How does bodily material become property in the first instance?

If one accepts the proposition "my body belongs to me," then I think there is a strong argument that extracorporeal bodily material should be considered, initially, the property of the person from whom it was derived. First of all, we can ask whether owning one's body means owning all of the constituent parts while they are still connected. I think the answer must be "yes." If I own a bicycle, then I own all of the nuts, bolts, chains, and wheels that comprise the bicycle. If I own a chair, then I own the legs, the seat, and the nuts and bolts that comprise it.

Does changing the location of bodily material from within my body to outside my body change my property rights in that material? I do not think so. If my chair is in my house and I take it out to my front lawn, then it is still my chair (although others will probably take steps to effect an illegitimate transfer of that chair if I do not take proper precautions). Of course, there are places that we can put property that, by convention, signal our intent to discard that property and relinquish our rights to it. For instance, if I put my chair in the dump and that chair later turned up in my neighbor's house, I could not accuse her of having stolen my chair.

In fact, pathologists have traditionally saved materials left over from operations, from blood samples given at hospitals, and so forth, and used that material without the consent or knowledge of the person from whom it was derived. They have done so under the assumption that a person who has an operation or gives blood discards his bodily material. Unfortunately, most people were not aware that operations or other medical procedures might yield scientifically useful material, and they were not aware that scientists or physicians were considering their cells and tissues as discarded and unclaimed.

Does it matter that an extracorporeal cell is transformed from being a part of the organism to a collection of cells in a culture dish? Probably not. If I owned a gold statue and melted that statue down to make jewelry out of it, the jewelry would belong to me. The right of ownership can include the right to waste or exhaust the things owned.

Would it matter that it was somebody else's labor that transformed my cells from being inside of my body to outside? Probably not. If I take my suit to the tailor to have it altered, her labor in transforming my suit does not make my suit hers. If I took my gold statue to a jeweler to have it melted down and turned into jewelry, the jewelry would still be mine, so long as I paid for the service performed.

I want to be clear about something that I am not proposing. I am not proposing that a person should own immortalized cell lines, cloned DNA, or other items that are created by scientists using human-derived bodily material as a starting point. I am proposing that a legitimate transfer such as a gift or sale from the patient or research subject to the scientist would have to take place before the researcher had the right to possess or use the patient's/subject's bodily materials for his own ends.

What if I believe that a person's rights in her body are not in the nature of property rights? Then could I still make the argument that cells taken from her body become her property if they are anybody's property? Or, would I be forced to contend that a person's cells, when taken from her body, are like unclaimed resources in some truly unexplored territory—owned by none and available for the taking (*res nullius*)? If I do not own my body, then is it true that I do not own its constituent parts? The arguments against owning bodies have to do with the fact that the self and the body are one, and that the self cannot be both subject and object at the same time. However, it is not clear that any particular organ (except perhaps my brain?), cell, or tissue is one with or identical to my self. So, perhaps arguments against owning bodies qua bodies do not apply to parts of bodies. Perhaps I don't own myself, but I could own individual parts.

Otherwise, the question of who owns the extracorporeal cells would depend on the theory (and law) of property acquisition one applied. Under some theories, such as a labor theory of property acquisition, the researcher or physician who removed the cells/tissues/organs and transformed them from their natural state into something more useful would probably have a greater claim to ownership than the person from whom they were derived.

Another alternative for one who does not believe that persons own themselves or their bodies is to argue that, regardless of ownership, we still have rights of privacy and exclusion in our bodies, and that those rights should also adhere to extracorporeal body parts.

COMMODIFICATION OF BODIES AND PERSONS

Acknowledging that one person could have property rights in another's bodily materials, or that I could have property rights in my own bodily materials, is not the same as arguing that we should do so. In fact, it really hasn't gotten us very far because we are still left with two questions. The first is what bundles of rights should various persons have in human body parts, former humans, and potential humans. The second questions is whether these bundles of rights should be conceived of as property rights or not. Just because certain metaphysical understandings of personhood do not rule out the possibility of property rights in human bodily materials doesn't mean that we should have them. Various rights of use, possession, and transfer can be assigned to achieve certain goals, including economic efficiency and scientific or medical progress, without necessarily denominating those as property rights.

To address my second question, whether a particular bundle of rights should be conceived of as property or not, we might want to know why it would matter whether the rights involved were explicitly designated as property rights. One reason it matters is that the body of law known as property law would then apply. If something is property, then the government may not deprive you of it without due process; it may not take your property without just compensation. Bodily materials conceived of as property would be subject to estate laws—they could be transferred by will to survivors, or they could be transferred by intestate succession. If the deceased had no survivors and no will, his bodily materials would revert to being property of the state. If bodily materials were property, they would be subject to marital property laws and split up like other marital property at divorce. These are just a few examples, but they serve to show that while labeling rights "property" does not fully define the parameters of ownership, affixing the property label would likely change the legal landscape with respect to how one's rights were respected and exercised.

A second and extremely important reason why it matters whether we designate a bundle of rights as property or not is because the words we use structure our thinking and our behavior. Different discourses have embedded within them different conceptions of self, others, the world, and its possibilities. We might feel that some discourses carry within them a high likelihood of transforming our social world in ways that we would find unacceptable. One problem with applying property rhetoric to human bodily material is that the buying and selling in the market is a defining conceptual feature of property. Market discourse exists within a market culture, a culture that advances and honors certain values that have typically been thought of as applicable to objects but not to persons.[18] Many fear that applying those values to human bodily materials will lead to a slippery slope in which those values are applied to persons. And further, that applying those values to persons would be inimical to human flourishing.[19]

What values of the market culture might be particularly threatening when applied to human bodily materials or human beings in general? There is a complex of related values that I think are problematic. These include commensurability, fungibility, and the reduction to or expression of all value as a monetary value.[20] "Commensurability" means that all things of value can be compared using the same scale or metric—the value of my child can be compared to the value of a car, a trip around the world, or a chance in the lottery. Commensurability is related to the issue of expressing all value as monetary, because money is the scale used to compare all valuable things. My child, my integrity, my love of my spouse, my car, my health—in a fully commodified world all of these are reducible to a dollar amount (and nothing more) for purposes of ranking them as being more or less valuable. Finally, "fungibility" means that things of equal monetary value are interchangeable with each other—one car that is worth $10,000 to me is interchangeable with $10,000 or with another car worth $10,000. If my child were worth $200,000 on the open child market, then I should be willing to exchange her plus $50,000 for a child worth $250,000. When this complex of values is applied to something, along with practices such as advertising, stockpiling, and arbitraging, I refer to that as "commodification."

Now, let us think about market practices and values as applied more fully to human embryos than they currently are. Would a certain amount of money mean the same thing to an individual, or satisfy her just as much, as an embryo created from her gametes? Would some other embryo, created from somebody else's gametes, ever be interchangeable with her own embryos (i.e., could there be an embryo that was worth the same amount of money to her as her own, and which would therefore be interchangeable with her own if she were operating solely according to marketplace values)?

If she created some embryos from her gametes and she were a good market participant, she might ask herself whether somebody else might value her embryos more than she does. For instance, perhaps she is a musician and out there in the world is a very successful but tone-deaf business person who would really like to have a child whose chances of having musical talent (including decent pitch) are high. So, she should advertise that she has embryos and see whether somebody will make her an offer she cannot refuse. Perhaps she should put up a Web site describing herself and the genetic father, the procedures that were used to make her embryos, and any genetic testing that has been done on the embryos. Maybe she should pay for some billboards or take out an ad in the *New York Times*.

What happens if she receives higher offers than any she expected? She sells off her embryos for a great deal of money. She decides to make more, to stockpile them under the assumption that the value of her embryos is going to increase. Then she starts a business teaching other people how to create valuable embryos; she recruits people to participate in her business who she believes have the biological and sociological characteristics that will cause their embryos to be highly valued in the market. Pretty soon she makes a deal with a major movie studio. When the studio releases a blockbuster movie with a big name star, instead of simply selling action figures in toy stores and McDonalds' meals, they can now auction off a few embryos made with that star's gametes!

She never has a child, but she becomes very rich. She cannot give her money a good night kiss, watch it take its first step, or enroll it in soccer. But under a purely market analysis, this would not matter because the money she got was worth it to her to give up these other things. Or, perhaps she used that money to buy things that satisfied her more than a child or some children would have.

And what about those people who choose not to sell their embryos? Do they think about their children and wonder what they could have brought in as embryos on the open market? Do they think about their own personal characteristics, including things like their weight, skin color, or athletic talents, and wonder how their personal characteristics would contribute to the worth of their embryos? Perhaps they would be concerned because nobody recruited them to make embryos. Perhaps they begin to wonder about their own value in the market.

This story suggests at least one mechanism by which a slippery slope could occur—we would move from applying market values to embryos to applying market values to their progenitors. Then, even those who do not enter the market might begin to rank themselves according to what their materials likely would have brought on the market had they entered it. What makes embryos, or sperm and eggs, valuable in the market is the qualities of the people from whom they derived. Are those people beautiful? Are they wealthy? Are they healthy? Are they great athletes? Are they Nobel Prize winners? These qualities are explicitly used in the selling of human reproductive cells. The same connection between the bodily materials and the person from whom they were taken may exist in other situations as well.

Of course, the slippery slope may never come to pass, or not in a manner that seriously impairs our human rights or human relations. Humans are amazingly complex creatures, capable of living with ambiguity and with conflicting or limiting beliefs and understandings. It is possible that we could live in a world in which some embryos were commodified and some were not, in which some human relations were commodified and some were not. In fact, we do live in such a world. Even though labor is commodified, some of us volunteer and do work that others of us are paid for. The fact that some are paid does not seem to diminish our experience as volunteers. As Professor Margaret Jane Radin points out in her book *Contested Commodities*, by automatically assuming that we will slide down the slippery slope, we are unnecessarily conceding power to market rhetoric.[21] We should not assume that this way of thinking is so seductive or satisfying that once presented it will necessarily become predominate in all spheres of our lives.

In fact, one thing that must be acknowledged is the degree to which there already exist markets in which human bodily materials are transferred.[22] There is a robust market for reproductive cells; sperm and eggs are sold on the Internet where their market value does largely stem from the qualities of the individuals from whom they derive. While programs that broker women's eggs claim not to be paying women for the eggs, but rather for their time and trouble (their services as egg donors), at least in some cases the women are not paid their full wage if they undergo the entire donation procedure and no mature eggs are recovered.[23] And although we technically do not pay for organs, organs are integral to a market for transplants in which large sums of money change hands. Although we do not technically pay for babies under surrogacy and adoption arrangements, sometimes the services and the "product" are not easily distinguishable.

Whether the exchange of money and the proliferation of market rhetoric, values, and behaviors surrounding these various exchanges of human bodily materials or reproductive activities will lead our society down a path in which we lose important and valued features of our human interactions and human rights is not yet a settled question. However, if our respect for each other diminishes, or our human relations are impoverished, by the time the trend becomes apparent, it may also be quite difficult to reverse.

CONCLUSION

I have argued that reasons justifying nonownership of human beings do not justify a refusal to allow property rights in extracorporeal human-derived bodily materials. There is a strong case to be made, however, that the individual from whom bodily materials are derived is the initial owner of those materials, and that legitimate transfers from them to scientists must take place before scientists can rightfully possess, use, or sell those materials, or exclude others from doing so. I have not taken a position as to whether further application of property concepts, market rhetoric, and market values to human bodily materials would be a good or bad thing. However, I have suggested at least one way in which a slippery slope could lead from the commodification of human bodily materials to the commodification of persons from whom those materials were derived and then to other persons as well.

One important point is that legal rules governing possession, use, and sale of human bodily materials can and should be promulgated, regardless of whether we affix the label "property" to these materials. In this essay I have not attempted to describe bundles of rights that should be attributed to bodily materials, in part because I think the answer varies depending on which materials and which markets you are regulating. Cells derived from tumors do not have the same uses or the same personal and legal significance as reproductive cells, and so different bundles of rights might reasonably be attached to extracorporeal human tumor cells and human eggs. Whether or not that is the case is beyond the scope of this essay.

NOTES

1. See UNOS, "Critical Data: U.S. Facts About Transplantation," <http://www.UNOS.org/Newsroom/critdata> (12 November 1999).

2. Rick Weiss, "Broader Stem Cell Research Backed," *Washington Post*, 11 September 2001, A1.

3. For decisions recognizing property rights in deceased humans see *Brotherton v. Cleveland*, 923 F 2d 477 (6th Cir 1991).

4. John Dwyer and Peter Menell, *Property Law and Policy: A Comparative Institutional Perspective* (New York: Foundation Press, 1998), vi.

5. John Harris, "Who Owns My Body," *Oxford Journal of Legal Studies* 16 (spring, 1996): 55.

6. For further discussion on this topic see, Pilar Ossorio, "Legal and Ethical Issues in Patenting Human DNA," in *A Companion to Genethics: Philosophy and the Genetic Revolution*, Justine Burley and John Harris (Oxford UK: Blackwell, 2002).

7. John Harris, "Who Owns My Body."

8. Carol M. Rose, *University of Chicago Law Review* 52 (1985): 73.

9. John Locke, *The Second treatise of Government*, ed. J.W. Gough (Oxford: Basil Blackwell, 1976), 27.

10. Immanuel Kant, *Lectures on Ethics*, trans. Louis Infield (New York: Harper & Row, 1961), 165.

11. Renee Descartes, *A Discourse of Method, Meditations on the First Philosophy*, trans. John Veitch, ed. A. D. Lindsay (Everyman Paperback Classics, Boston: Tuttle, 1994.)

12. For various views on personal identity see John Perry, *Personal Identity* (Berkeley: University of California Press, 1975); see also Derek Parfit, *Reasons and Persons* (New York: Oxford University Press, 1984).

13. *McFall* v. *Shimp*, 10 Pa D & C 3d 90 (1978).

14. For an example of a case where a woman was compelled to undergo a Cesarean section see *Jefferson* v. *Griffin Spalding County Hospital Authority*, 274 SE 2D 457 (Ga 1981).

15. *Bauer et al* v. *North Fulton Medical Center, Inc.*, 241 Ga App 568, 571 (1999).

16. Margaret Jane Radin, "Property and Personhood," *Stanford Law Review* (1982): 957, 959.

17. Meir Dan-Cohen, "The Value of Ownership," *Global Jurist Frontiers* 1, no. 2 (2001): article 4.

18. Margaret Jane Radin, *Contested Commodities: The Trouble With Trade in Sex, Children, Body Parts, and Other Things* (Cambridge, Mass.: Harvard University Press, 1996).

19. Ibid.

20. Ibid.

21. For further discussion of this point see Ibid., Chaps. 7 and 11.

22. For further discussion of this point see Julia D. Maloney, "The Market for Human Tissue," *Virginia Law Review* 86 (2000): 163.

23. Ruth Macklin, "What Is Wrong with Commodification?" in *New Ways of Making Babies: The Case of Egg Donation*, ed. Cynthia Cohen (Bloomington, Ill.: Indiana University Press, 1996).

twelve

ETHICAL ISSUES IN THE PATENTING AND CONTROL OF STEM CELL RESEARCH

GLENN MCGEE AND ELIZABETH BANGER

H uman embryonic stem cell (hES) research has been the focus of intense public discussion since it was announced in 1998 that stem cells had been isolated from human embryos in the labs of Johns Hopkins University and the University of Wisconsin.[1] Long before any clinical demonstration that embryonic stem cells could have therapeutic efficacy in the treatment of human disease, many scientists and politicians were willing to speak and write of "the profound potential" of stem cells for medicine.[2] Those who objected to abortion, fetal tissue research, and/or in vitro fertilization on moral grounds were equally quick to condemn embryonic stem cell research in the strongest possible terms, advocating instead the use of stem cells derived either from adults or from blood obtained from the umbilical cord.[3] The scientific facts that would make clear the advantages of proceeding with embryonic or adult approaches are not yet in evidence, yet the reliance of both pro- and anti-hES research groups on scientific arguments has left many in a state of confusion about how the facts of the matter relate to the underlying moral concerns. Both sides have sought middle ground.

One attempt to resolve the debate over stem cell research involves the suggestion that researchers might obtain stem cells from embryos without actually engaging in the destruction of those embryos.[4] In just such an effort, U.S. President George W. Bush suggested that while it is in his view immoral to destroy embryos, some accessible hES cells have been derived from embryos that have already been destroyed—and the matter of the availability of those cells can be considered distinct from the matter of creating new cells through the destruction of additional embryos. He thus decreed that only stem cells derived from embryos destroyed prior to his speech would be made available for federal funding. As the president framed his compromise, "only those cells for which the life or death decision has already been made" would be eligible for use.[5] He noted that sixty-six stem cell lines have already been obtained from embryos, "more than enough" to allow that research to proceed.

Predictably, a number of concerns were raised about the president's rationale and about his policy. However, the overriding question was whether enough embryonic stem cells, in fact, exist. The issue of the suitability and scarcity of hES cell lines already derived at the time of President Bush's speech called attention to the fact that many human embryonic stem cell lines are subject to U.S. and international patents, and that many of the innovations necessary to derive, culture, differentiate, or otherwise manipulate stem cells are also subject to patent.[6] But *should* stem cells, embryos, embryo-like organisms, or the cells derived from them be eligible for consideration as intellectual property, whether through patents or other protections of law? Patient advocacy groups and others, including the Appropriations Committee of the U.S. Senate, have expressed concern that stem cell research may be hindered, and that eventual stem cell therapies may be subject to excessive levies by patent holders unless the matter of stem cell patenting is taken up by the courts and other relevant governmental bodies.[7]

Following a brief introduction to stem cell research, it is our purpose in this essay to examine only those aspects of the ethical debate about stem cell research that bear on the broader question of when and whether governments should allow the patenting of technologies that derive wholly or in part from examination of the mammalian embryo. We examine representative, important, and recently issued U.S. patents in the area of human and animal stem cell research, particularly those patents that concern research on embryonic stem cells (hES). We discuss three problems with these patents: (1) the difficulty of determining what the patents actually claim or whether the claims of a patent are so broad or diffuse as to suggest overlap with other patents; (2) the potential ethical implications of allowing patents on the specific innovations claimed by these patents; and (3) the potential implications of these patents for the conduct of basic and clinical research in stem cell biology.

★★★

From the point of view of consumers, activists, and patients, stem cell research may seem to have materialized from nowhere, a miraculous discovery with great potential.[8] Yet stem cells have long figured prominently in basic research in human and veterinary cell biology, in clinical trials of possible therapeutic techniques, and even in a number of successful therapies.[9] Basic research involving stem cells is most often focused on fundamental problems of developmental biology, for example, how it is that specialized cells come into being, and how groups of specializing cells come to participate in coordinated activities.[10] Basic stem cell research thus focuses on the time in, manner through, and extent to which somatic cells specialize during the development of an organism, and the role of stem cells in repopulation and repair of damaged or otherwise depleted cells in the mature organism. Several well-publicized clinical trials involved the transplantation of fetal tissue into patients with degenerative diseases of the brain and nervous system, such as Parkinson's disease. While these trials did not specifically measure the activity of stem cells, they raised basic questions about the utility and toxicity of immature cells for transplantation. Clinical research that specifically involves stem cells has included a wide variety of tests of the effectiveness of transplanted stem cells in repopulating certain needed cell types in patients with, for example, bone cancer and diseases of the immune system.[11] Techniques already in use include the harvesting of stem cells from umbilical cord blood and the transplantation of stem cells for the treatment of leukemia.

John Gearhart and James Thomson, alongside perhaps one hundred other senior and junior researchers at Johns Hopkins, the University of Wisconsin, and a dozen other universities, had raced to identify the *pluripotent* human embryonic stem cell for several years prior to the publication of their findings in 1998. The Hopkins and Wisconsin groups, along with many others in the field, worked in an interdisciplinary and unusually imaginative context in which stem cell research was fashioned out of a variety of disparate research interests; unlike much of contemporary genomics, which has become very much goal directed and focused in character, the labs of stem cell research had dozens of possible trajectories in mind for their work.

While it was not clear at the time of Gearhart and Thomson's publications exactly what would result from the identification and cultivation of pluripotent hES cells, it was immediately apparent that their findings had great importance for both basic and clinical research in humans and animals. First, Gearhart and Thomson had identified a key point in the development of the human embryo at which the DNA in the nucleus of particular, undifferentiated cells no longer has the power to make another identical organism—the point at which *totipotency* is definitely not present. Second, and more important, these cells' nuclei can produce a wide range, and perhaps all, of the kinds of cells that populate a developing or mature human organism. Third, it is possible to derive these cells from the embryo and to isolate them from other cells. Fourth, once derived, these isolated pluripotent human embryonic stem cells can be cultured and frozen, transported and grown, fed and measured in a variety of ways.[Fifth, these cells can be induced to produce differentiated cells, cells that might then themselves produce more cells and which might be transferred from culture into the bodies of patients to replace a wide variety of damaged cells, or to perform a range of other tasks, from inoculation to the destruction of cancerous tissues to the delivery of drugs.]

Enthusiasm about embryonic stem cell research quickly led to a larger discussion of the future of the work and the implications of stem cells for broader debates about how to allocate healthcare resources, how to proceed with caution in new areas of clinical research, and how to regulate research involving embryos, fetuses, or abortion. Wide calls for governmental investment in stem cell research were entertained both as part of the 2000 presidential campaign in the United States and as part of governmental hearings the world over. It was noted in the United States and elsewhere that, like mammalian cloning research, most innovation in stem cell research was being made by researchers whose work was funded by small companies rather than national or regional governments.[Arguments for government funding of stem cell research were almost always linked to the claim that government funding would enable regulation, and if necessary restriction, on stem cell research. This argument received the endorsement of many ethics advisory boards, including, for example, the U.S. National Bioethics Advisory Board (NBAC).[12] What did not emerge immediately was the question of how patents filed in association with Gearhart, Thomson, and others might make it difficult for the government to exercise as much regulatory authority or research leadership as was sought by groups like the NBAC, the American Association for the Advancement of Science (AAAS) and others.

The effect and potential effect of patents on the conduct of stem cell research or stem cell therapies has become a timely matter in the United

States. In 2001 a presidential order authorizing National Institutes of Health (NIH) funding for embryonic stem cell research—but only on cells derived from embryos already destroyed at the time of his speech, and then only where the embryos had been created for reproductive purposes—was discarded because it was no longer needed for those purposes, and for which no compensation was given to the donors of the embryo. President Bush announced a registry, which in the United States is called the Human Embryonic Stem Cell Registry, that lists the hES cell lines that meet the eligibility criteria. As of November 15, 2001, seventy-two cell lines had been declared on that registry's Internet site, http://escr.nih.gov/, of which twenty-seven were held by labs or companies in the United States, six by an Australian company, twenty-five by Swedish universities, ten by Indian research centers, and four by an Israeli university. Of the twenty-seven cell lines in the United States, twelve are held by or with Geron Incorporated of Menlo Park, California, or the University of Wisconsin Alumni Research Foundation (WARF) of Madison, Wisconsin, who hold patents relevant to hES research generally and to their cell lines specifically. The remaining fifteen are held by BresaGen Incorporated of Athens, Georgia (four), CyThera Incorporated of San Diego, California (nine), and the University of California at San Francisco (two). In a search of published and Internet materials of these three organizations and of the online record of U.S. patents, we were not able to identify any hES-revelant patents held by the organizations who hold the fifteen U.S. cell lines not produced by Geron or WARF.[13]

There are significant questions to be answered about how many of the seventy-two hES cell lines in the NIH registry are actually viable for research, and many have alleged that only a fraction will, in fact, be useful due to contamination or an advanced stage of cellular development. Some cells may have been derived from embryos with maladies that would prevent proper cellular development or may not have been properly isolated from other cells in the embryo. Among all the hES cell lines, those best characterized and most manipulated are held by Geron and WARF, the organizations associated with the labs of Gearhart and Thomson. These cells are certainly subject to patent. In addition, though, recent litigation by Geron and WARF against each other suggests that each believes that it has patent rights to control not only each other's cells but also many if not all of the sixty cell lines in the registry that were developed by others. As debate has ensued about how many embryonic stem cell lines actually exist, how many were derived prior to the date of the announcement, and how many are of sufficient quality for use in research in the United States and elsewhere, the attention of scientists and policymakers has further turned to the fact that many and perhaps all human embryonic stem cells yet to be derived and/or made may already be subject to the patents developed by those working on stem cell research in the few years since 1998. The fear of restrictions on the availability of hES cells, either due to governmental restrictions or due to patents, began to draw the attention of the U.S. Senate, where committee hearings were held on the effect of intellectual property protections on stem cell research.[14]

PATENTS OF CONCERN

In order to understand the extent of intellectual property in stem cell research, or to speculate about the potential impact of patents in that research area, it is necessary to conduct an inquiry into existing patent claims on stem cell research.

To that end, we identified the names of all researchers holding either an M.D. or a Ph.D. degree (or both) who are listed on the Web sites of companies (1) that conduct research on stem cells, and (2) that are in addition traded on the U.S. NASDAQ stock exchange. In addition, we identified the names of researchers at two organizations that do not trade on the NASDAQ, Advanced Cell Technology and WARF. We also identified the names of researchers in institutions who have published articles in the area of hES cell research since January of 1999. We then correlated the resulting list of investigator names with a list of patents issued after that date that include the phrase "stem cell(s)," and read those patents' abstracts to determine whether (1) any of the researchers in our list or their affiliated organizations were involved in the patent, and (2) whether the patent directly concerns stem cell research or merely involves stem cells in some ancillary way. More than one hundred patents were issued between January 1999 and October 2001 that involve stem cells, of which most focus on the stem cell or its activity; as noted in table 1, we identified nineteen patents that are both directly focused on stem cells and issued to assignees associated with, or actual investigators from our list. In table 1 each patent

TABLE 1: PATENTS IN THE AREA OF STEM CELL RESEARCH BY CATEGORY (as of November 15, 2001)

PATENT NUMBER	INVENTOR	ASSIGNEE	DATE ISSUED	TITLE	TOPIC
HEMATOPOIETIC STEM CELLS					
6,255,112	Thiede and Mbalaviel	Osiris Therapeutics	3-Jul-01	Regulation of hematopoietic stem cell differentiation by use of human mesenchymal stem cells	inducing hematopoietic stem cells to differentiate into osteoclasts; can be genetically engineered
6,280,718	Kaufman and Thomson	Wisconsin Alumni Research Foundation	28-Aug-01	Hematopoietic differerentiation of human pluripotent embryonic stem cells	methods of obtaining hematopoietic cells from cells using mammalian stromal; cells derived this way are good for transplants, transfusions, etc.
METHODOLOGY AND STEM CELLS					
5,843,780	Thomson	Wisconsin Alumni Research Foundation	1-Dec-98	Primate embryonic stem cells	purified preparation of primate embryonic stem cells with specific characteristics and the resultant stem cell lines
6,090,622	Gearhart and Shamblott	Johns Hopkins School of Medicine	18-Jul-00	Human embryonic pluripotent germ cells	germ cells extracted from post-blastocyst embryos and their culture and genetic manipulation
6,110,739	Keller, Hawley, Choi	National Jewish Medical and Research	29-Aug-00	Method of produce novel embryonic cell populations	novel immortalized precursor cell and populations derived from embryonic stem cells and methods for producing these cells
6,194,635	Anderson and Shim	The Regents of the University of California	27-Feb-01	Embryonic germ cells, method for making same, and using the cells to produce a chimeric porcine	method of producing a sustained culture of pluripotent porcine embryonic germ cells with specific characteristics
6,200,806	Thomson	Wisconsin Alumni Research Foundation	13-Mar-01	Primate embryonic stem cells	purified preparation of primate (including human) embryonic stem cells, proliferation in an undifferentiated state, and method for isolating stem cell lines

PATENT NUMBER	INVENTOR	ASSIGNEE	DATE ISSUED	TITLE	TOPIC
6,245,566	Gearhart and Shamblott	Johns Hopkins School of Medicine	12-Jun-01	Human embryonic germ cell line methods of use	primordial germ cells isolated from and human embryonic tissue are cultured to get cells that resemble embryonic stem cells; can be cultured and genetically engineered
6,251,671	Hogan and Zhao	Vanderbilt University	26-Jun-01	Compositions and methods of making embryonic stem cells	cell proliferation, cell differentiation, male/female infertility and compositions and methods
ANIMAL PATENTS					
5,843,780	Thomson	Wisconsin Alumni Research Foundation	1-Dec-98	Primate embryonic stem cells	purified preparation of primate embryonic stem cells with specific characteristics and the resultant stem cell lines
5,945,577	Stice, Cibelli, Robl, Golueke, Ponce de Leon, and Jerry	UMass at Amherst	31-Aug-99	Cloning using donor nuclei from proliferating somantic cells	an improved method of nuclear transfer involving transplanting differentiated cell nuclei from a donor cell into an enucleated oocyte as it relates to the manipulation of genetic structure and the production of various types of cells, including stem cells
5,994,619	Stice, Cibelli, Robl, Golueke, Ponce de Leon, and Jerry	UMass at Amherst	30-Nov-99	Production of a chimeric bovine or porcine animals using cultured inner cell mass cells	claims a method for producing a chimeric bovine or chimeric porcine using cultured inner cell mass (CICM) cells
6,011,197	Strelchenko, Betthauser, Jurgella, Pace, and Bishop	Infigen, Inc.	4-Jan-00	Method of cloning bovines using reprogrammed non-embryonic bovine cells	culturing cells from a bovine fetus (35–70 days gestation) and using nuclear transfer to enucleate a bovine oocyte
6,103,523	Moreadith and Schoonjans	Thromb-XN.V (Belgium)	15-Aug-00	Pluripotent rabbit cell lines and method of making	claims pluripotent rabbit cell lines with specific characteristics, a method for producing those lines, and a selective medium used in the process

PATENT NUMBER	INVENTOR	ASSIGNEE	DATE ISSUED	TITLE	TOPIC
6,107,543	Sims and First	Infigen, Inc.	22-Aug-00	Culture of totipotent embryonic inner cells mass cells and production of bovine animals	claims a method for producing a bovine using embryonic inner cell mass cells that have resulted from several different methods
6,147,276	Campbell and Wilmut	Roslin Instit., Minister of Agriculture, Fisheries and Food, Biotechnology and Biological Sciences	14-Nov-00	Quiescent cell populations for nuclear transfer in the production of non-human mammals and non-human mammalian embryos	method for reconstituting a non-human mammalian embryo using specific techniques and the resultant off spring from those embryos
6,258,998	Damiani, Betthauser, Forsberg, and Bishop	Infigen, Inc.	10-Jul-01	Method of cloning porcine animals	method for cloning pigs using totipotent cells, porcine embryos, nuclear transfer, and claims on the resulting animals
6,281,408	Khillan	Thomas Jefferson University	28-Aug-01	Efficient method for production of compound transgenic animals	producing trangenic animals by co-culturing embryonic stem cells (specifically chimeric mice)
6,287,863	Hodgson	Nature Technology Corporation	11-Sep-01	Method of transferring a DNA sequence to a cell in vitro	using vectors to transfer foreign DNA to promote expression in genes as a means of therapy

is identified by title, assignee, relevant scientist, and topical area. Banger's descriptions of the topical areas are drawn both from the description section of each patent as well as from each patent claim's detail.

Limitations of our strategy include (1) the fact that we were not able to identify patent applications still pending, (2) our failure to correlate issued patents with specific journal articles disclosing the findings enumerated in the patent, and (3) the weakness of our reliance on the key words "stem cell(s)," which while probably sufficient to identify all patents of immediate relevance to the issues as we describe them may well have missed patents whose authors elected to describe their cells without using the term "stem." Nonetheless, this first review of existing U.S. claims for protection of novel intellectual property in the field of stem cell research yielded interesting results that suggest a need for both a more thorough review and a discussion of ethical issues involved in the patents already issued.

Patents on stem cells can be divided into four main categories, each of which corresponds to the kind of activity studied or produced by the researchers: (1) patents that deal with hematopoietic stem cells, for example, Kaufman and Thomson's patent on "hematopoietic differentiation of human pluripotent embryonic stem cells" and Thiede and Mbalaviele's patent on "regulation of hematopoietic stem cell differentiation by use of human menchymal stem cells"; (2) patents that relate to specific manipulations of stem cells, for example, Yale's patent employing the notch pathway; (3) patents that deal with methodology and stem cells, for example, Vanderbilt's patent on spermatagonial stem cells; and, (4) patents that relate human and animal stem cells, either through the use of mammalian cloning technology or through other derivation and/or development of animal stem cells for human use. Despite the differences among the nineteen patents in these four categories, more is shared than not. Apart from those patents that deal with inducing embryonic stem cells to produce a *specific* kind of somatic cells, all of the patents not focused on nuclear transfer share a great deal in common; among those patents held by Geron, WARF, and the researchers associated with Advanced Cell Technology (those whose assignees include the University of Massachusetts), even more is shared. It is significant that so many similar patents with great potential for overlap and disputes over infringement were issued in the first place, but it is *what* is shared that is of greatest importance. These are in the main patents that claim the most basic aspects of human embryonic stem cell activity; many make claims that are as much about observing the typical activities of the embryonic cell as about innovating through the use or modification of the cells or their activities.

Both concerns are most apparent in the dispute between Geron and WARF concerning the claim made by each that it has the rights to the other's cells and future research. Each of these organizations' patents are astonishingly broad; the widest claim made by Gearhart (whose assignee is Johns Hopkins University, and who is associated with Geron Inc.) is that in U.S. Patent No. 6,090,622 to "human pluripotent embryonic cells, wherein the cells exhibit the following culture characteristics during maintenance." Subsidiary claims identify the many specific characteristics that identify the cells he claims, including virtually all imaginable characteristics of hES cells, for example, "wherein the growth factor is basic fibroblast growth factor." The broadest claim made by inventor Thomson (of the University of Wisconsin, and whose assignee is WARF) is found in U.S. Patent No. 6,200,806 to "a purified preparation of pluripotent human embryonic stem cells which will (1) proliferate in an *in vitro* culture for over one year, (2) maintains a karyotype in which the chromosomes are euploid and not altered through prolonged culture, (3) maintains the potential to differentiate to derivatives of endoderm, mesoderm, and ectoderm tissues throughout the culture, and (4) is inhibited from differentiation when cultured on a fibroblast feeder layer."

These patents are virtually identical in terms both of what is described and how it is to be identified and cultivated according to the claims in each patent. Gearhart and Thomson make claims using the same set of markers to identify their cells. SSEA-1, SSEA-4, TRA-1-60, TRA-1-81, and alkaline phosphatase activity are these factors. For Thomson, the markers are SSEA-1 negative, SSEA-4 positive, TRA-1-60 positive, TRA-1-81 positive, and positive for alkaline phosphatase activity. For Gearhart, the markers are as follows: SSEA-1 positive, SSEA-4 positive, TRA-1-60 positive, TRA-1-81 positive, and positive for alkaline phosphatase activity. The distinction between these two patents is minimal even at the level of fundamental observations of the cells' characteristics.[15] Whether the patents overlap *in toto* is less important than the fact that both Gearhart and Thomson's claims overlap to some extent all seventeen other patents in table one. Moreover, twelve of the seventeen other patents overlap at least three other patents plus those of Gearhart and Thomson. It is not at all obvious how the U.S. government determined that each individual patent was, in fact, eligible given other patents already issued or in submission. The groundwork is clearly laid for an all-out legal battle to determine whether the U.S. government erred in issuing overlapping patents and, more important, a battle for entitlement to the intellectual property claimed in the overlapping regions.

Whether or not the patents overlap, each—and particularly those of Thomson and Gearhart—makes extraordinarily broad claims to many aspects of hES cells as they are discovered in their embryonic environment. Take, for example, Thomson's claim #9: "A method of isolating a human embryonic stem cell line, comprising the steps of:

(a) isolating a human blastocyst
(b) isolating cells from the inner cell mass of the blastocyte of (a)
(c) plating the inner cell mass cells on the embryonic fibroblasts, wherein inner cell mass-derived cell masses are formed.
(d) dissociating the mass into dissociated cells
(e) replating the dissociated cells on embryonic feeder cells
(f) selecting colonies with compact morphologies and cells with high nucleus to cytoplasm ratios and prominent nucleoli; and
(g) culturing the cells of the selected colonies to thereby obtain an isolated pluripotent human embryonic stem cell line."

In this claim, Thomson describes the ways in which he finds and grows cells that always already exist in the typical in vitro embryo. He then claims himself as inventor not only of the stem cells themselves, but the resulting cell lines, in essence a claim to whatever use is made of those cells in virtue of his claim to the derivation of the cells in the first place. It is a claim that reads very much like a description of the method involved in digging for gold, but which then claims the gold itself as an invention. The digging is supposed to have "purified" the gold, but the cells are valuable exactly because they are able to behave as they would ordinarily do in one, or several, of their ordinary roles in the developing embryo or in the mature organism that would ordinarily result from embryonic and fetal development.

Not surprisingly, neither WARF nor Geron is happy with the broad claims of the other. WARF filed a lawsuit alleging that Geron should not be granted exclusive rights to any additional stem cell types it develops, because it has not demonstrated the utility of the cells it claims in its patents, or in particular the utility of the cells it received an exclusive right from WARF to develop in 1999, after Geron helped fund the hES isolation work of University of Wisconsin research by James Thomson. It, moreover, has become a matter of controversy that because both WARF's and Geron's patents are so broad as to conceivably cover all stem cells derived in the future from embryos, or even their products, both organizations might be entitled to levy fees, exert restrictions, and collect some share of profits on eventual outcomes of research funded by the U.S. NIH funding initiative in stem cell research. Secretary of Health and Human Services Tommy Thompson, former governor of Wisconsin, was quick to point out that he believed that WARF would not charge its typical $5,000 per line fee on the use of stem cells to those researchers funded by the new government initiative, and WARF itself announced with the NIH that it would not ask for a share of profits obtained from novel research conducted using its stem cell lines.

However, both Gearhart and many other critics including coauthor McGee alleged that it was not obvious, from WARF's declaration, that WARF will in fact rescind all rights to commercial developments involving any use of its cells by federally funded researchers. Moreover, the conflict of interest involved in Secretary Thompson's speaking, negotiating, and executing documents of policy about intellectual property so clearly tied to profits for an organization in (and part of the governmental structure of) his home state struck McGee as significant enough to warrant at least some public discussion if not an active effort to mitigate the conflict.[16] It is enough to note the mere fact that the focus of concern is the most powerful person in the structure of U.S. government funding of medical research, a person most governmentally funded researchers—and certainly many researchers without a financially protective tenured professorship—will find very difficult to criticize on the matter.

ARE STEM CELLS PRODUCTS OF NATURE OR INNOVATION?

Among those problems of patenting in molecular biology, the most intractable and metaphysically complex is surely the fundamental question of

whether the matters claimed by many gene patents are innovations of investigators or observations of natural phenomena. As Rebecca Eisenberg has
noted, the standard response to the problem of patenting DNA sequences as
compositions of matter has been to draft the language of a patent application in such a way as to claim that the identification and use of claimed
sequences of DNA required the "isolation and purification" of that DNA
(through removal of the DNA from an organism and a chromosome, and
potentially through introduction of the DNA into recombinant vectors).
The patent holder is then able to exercise control over the use of the composition of matter, but not to restrict others from analyzing information
about what that sequence is and does. As Eisenberg points out, the lawyerly
technique of drafting patents in this way serves not only the interest of the
applicant but also the general public interest in distinguishing between the
activities of cells or molecules in the body, which ought not be protected by
patents, and the use of those cells or molecules in a lab setting.

However, many new patents in molecular biology involve the capture
or attempted capture of patent protection not only of the composition of
matter (the DNA itself, "purified and isolated"), but also of the information to be gleaned from that composition of matter, including not only
information related to the activity of that DNA in its natural setting, but
also to the information that comprises the sequence of the DNA when
laid out as a stream of data.

In just this way, stem cells that occur "in nature" as that term is understood by the U.S. Patent and Trade Office (PTO) are, in all nineteen
patents listed in table 1, "isolated" (removed) from the in vitro embryo and
induced to perform ordinary activities in the setting of the lab. While it is
a matter of legal standing as to whether the description of how these cells
are moved or cultivated is sufficient to pass muster with the PTO or with
the courts and agencies of the United States and other nations, it is a much
broader and more difficult matter to assess whether in fact it is reasonable
to argue that the derivation and cultivation of stem cells and cell lines
qualifies as the sort of activity best understood as innovation—and more
important, whether the cells and cell lines themselves can thus be protected
as a product and as part of that innovation, whether through patent protection of the information that describes those cells and cell lines and/or
through protection of the products of those cells and cell lines. While
DNA and other molecular genetics patents have expanded to include these

latter informational claims, stem cell patents have begun with both composition of matter and informational elements.

The broad language of stem cell patents, including both the composition and the information defining and produced by the cells and their products, is significant in part because it follows on the heels of gene patenting, and at a time when the rate of submissions of patents to the PTO is increasing exponentially. But the more important fact about these patents is that they are so broad at such an early stage of the research. Patents on stem cell technology, and particularly those associated with the investigators whom we identified in table 1 and those associated with the Bush administration–approved cell lines (as identified in table 2), arguably give controlling authority for stem cell research to small institutions and groups of investigators, and to corporate control, at a time when the science of stem cells is quite early. In fact, judging by the published quarterly reports of those companies engaged in hES research who trade on the NASDAQ exchange, it is exactly the control over cell lines and their products that is to be the primary source of value for companies working in this area.

As opposed to molecular genetics, where a mature basic science community exists and the threat of excessive patent domain runs into conflict with the sensibilities of thousands of scientists working in relevant areas around the world, the area of stem cell research is relatively small and is peopled in the main by scientists who are associated with organizations very much like those assigned patents in table 1. The entire field of stem cell research (both in basic and clinical science) is still very much emergent, and yet patent protection is already in place that could allow a tiny number of companies to exert enormous influence over the conduct of stem cell research for many years. This is so because, as Rebecca Eisenberg points out in chapter 6, informational patents are "particularly ill suited to the protection of information because there are so few safety valves built into the patent system that constrain the rights of patent holders in favor of competing interests of the public. Unlike [U.S.] copyright law, patent law has no fair use defense that permits socially valuable uses to go forward without a license."[17] While Eisenberg has in mind the applicability of patent protection to genetic material, the point is more germaine to stem cell research, where broad patents framed in terms of compositions of matter *and* information have already issued, potentially preventing labs with interest in stem cell research from engaging in even basic work on

TABLE 2: PATENTS RELATED TO THOSE ORGANIZATIONS IDENTIFIED BY THE U.S. NATIONAL INSTITUTES OF HEALTH AS HAVING CELL LINES AVAILABLE FOR RESEARCH UNDER U.S. FEDERAL FUNDING (as of November 15, 2001)

PATENT NUMBER	INVENTOR (n=4)	ASSIGNEE (n=2)	DATE ISSUED	TITLE OF PATENT	STEM CELL LINES REGISTERED WITH NIH BY ASSIGNEE- OR INVENTOR-AFFILIATED ORGANIZATION (n=24)
6,090,622	Gearhart and Shamblott	Johns Hopkins School of Medicine	18-Jul-00	Human embryonic pluripotent germ cells	7 registered under Geron
6,200,806	Thomson	Wisconsin Alumni Research Foundation	13-Mar-01	Primate embryonic stem cells	5 registered under WARF
6,245,566	Gearhart and Shamblott	Johns Hopkins School of Medicine	12-Jun-01	Human embryonic germ cell line and methods of use	7 registered under Geron
6,280,718	Kaufman and Thomson	Wisconsin Alumni Research Foundation	28-Aug-01	Hematopoietic differentiation of human pluripotent embryonic stem cells	5 registered under WARF

hES cells without first paying a license (in many cases) and then allowing the patent holder to negotiate for a share (and potentially for all) revenue that might eventually be derived from the use of the cells or products.

The concern, then, is not so much about the metaphysical question of whether stem cells as claimed in table 1 are a phenomenon of nature, as it is a question of what it will mean to allow such patents to stand at such an early stage of stem cell research. It is easy to see how the research came to this juncture: governments have long refused to fund reproductive research including that on or involving embryos or their cells, and as a consequence scientists with interest in such matters have turned to venture capital. Success in early stem cell research coincided nicely with the evolution of a business model in the computer and information sciences industry of financing small efforts whose early product would be intellectual property that allows the small research group to leverage its control over basic technology into capital for future research—and allows the venture capital lenders to assert the production of a product even where no actual consumer-ready item exists. Yet the clear path to the present dilemma does not diminish its urgency. As the Federal Court of Appeals 7th Circuit held in *Dickey-John Corp.* v. *International Tapetronics Corporation* in 1983, "[O]ne cannot patent the very discoveries which make the greatest contributions to human knowledge, such as Einstein's discovery of the photoelectric effect, nor has it ever been considered that the lure of commercial reward provided by a patent was needed to encourage such contributions. Patent law's domain has always been the application of the great discoveries of the human intellect to the mundane problems of everyday existence."[18] If one cedes that patents on human stem cell lines and the methods for finding them are merely a matter of applying great discoveries, where then are the great discoveries about stem cells open to the domain of broad investigation by scientists and the public? It would seem arguable that patents in stem cell research begin so early in the research, and at such a fundamental level, as to not only impede but literally give control over the most basic areas of stem cell research to a small group of feuding entrepreneurs—at a time of both great promise and enormous controversy.

WHAT EFFECT WILL
STEM CELL PATENTS HAVE ON THE
REVIEW AND CONDUCT OF RESEARCH?

Scientists and politicians are not the only ones interested in problems of access to stem cells and stem cell research. It was argued elsewhere by coauthor McGee, in resigning from the ethics advisory board of one of the three largest stem cell groups, that the private funding of much of stem cell research meant that organizations had little incentive to open their labs to peers, the media, or ethical scrutiny.[19] To date, the only real access to company activities in stem cell research is through voluntary disclosures such as those offered in scientific or business publications, through the filing of patents, and through mandatory disclosure to stockholders of events or activities that might materially affect the value of the company's stock. However, in the latter two cases, much confusion remains about when and what disclosure *must* be made, and by whom. Are privately held companies with no governmental funding under any legal obligation to disclose their relationship to patent-holding scientists working with the company? How much information must they reveal about the breadth of their patent portfolio? Do privately held companies have any requirement to disclose their involvement in controversial basic research prior to, or subsequent to, the company's filing of a patent? Answers to such questions have not been forthcoming, but there is increasing urgency attached to them in the media as a result of several discoveries about corporate research not known to peers of those conducting the research.

When, for example, the Jones Institute in Virginia announced that it planned to create new stem cell lines by producing embryos for use in research, it was not long before a *Wall Street Journal* reporter noticed that a company, Advance Cell Technology (ACT), had already begun its own effort to do something very similar, but had kept its own efforts secret.[20] Confronted with the question of whether it was engaged in controversial research without adequate peer review or public debate, ACT pointed to the fact that it had arrived at consensus about the ethics of its procedure in an internally sponsored ethics advisory board (EAB). Jones Institute, too, pointed to an ethics advisory board as an advisory, if not governing, mechanism. But neither organization would name the members of its EAB, or open its files or plans to other ethicists or the media. Advance Cell Technology, which does not trade shares of stock on a public exchange, is not required to make disclosures or offer access, nor is Jones Institute, which like ACT does not receive governmental funding for the research in question. Moreover, because these organizations can keep proprietary the nature and activities of their research from competitors and the media for long periods of time, study of ethical issues involved with stem cell research may be limited either to that which is possible through membership on a corporate ethics board (operating under rules of secrecy), or to research based on educated guesses about what is going on in the lab. The same limits apply to governmental review and policy promulgation: until an experiment falls under the jurisdiction of the FDA or other governmental authority, either through the very limited funding mechanisms created by the United States and other nations or through oversight of research involving hazardous substances or human subjects, the governing body will be able to do little if anything about regulating the overall endeavor. Just as the United States has been unable to really regulate genetic testing in the absence of voluntary applications for FDA approval of genetic tests, it will be extraordinarily difficult—short of the creation of a special oversight body for stem cell research, which would be difficult to justify as a constitutional restriction on research unless there is funding for all studies to be reviewed by the body.

It is difficult to know what implications stem cell patents will have for the conduct or cost of clinical trials or therapy using derivative technologies. It may be months or years before hES research yields therapies ready for human clinical trials, and still longer before it is clear what research might have been initiated if there had not been excessive cost or restriction attached. It is, however, clear that patents will have an effect both in the short-term development of research programs under U.S. funding and in the longer term as the research area evolves. The value of stem cell research is literally, and from its origins, quantified in dollars, and universities (apart from the half dozen with strong patents in the area) may not play as large a role in the basic research in stem cells as would have been the case without patent protections, or may have to spend as much on lawyers and patent licensing as on bricks and mortar or faculty recruitment. Small companies may elect to sell their patents, or even their entire research endeavor, to larger life sciences companies or to pharmaceutical corporations, which have thus far been reluctant to talk about investing in hES research. And some stem cell research and education programs will no doubt never begin, or never achieve leadership, as a result of the new paradigm in which whole categories of basic research are open from their very beginnings to protection as intellectual property.

NOTES

1. James A. Thomson et al., "Embryonic Stem Cell Lines Derived from Human Blastocysts." *Science* 282 (6 November 1998): 1145–47.

2. Paul Wolpe and Glenn McGee, "'Expert Bioethics' as Professional Discourse: The Case of Stem Cells," in *The Human Embryonic Stem Cell Debate*, ed. S. Holland et al. (Cambridge: MIT Press, 2001).

3. Richard M. Doerflinger, "The Ethics of Funding Embryonic Stem Cell Research: A Catholic Viewpoint," *Kennedy Institute of Ethics Journal* 9, no. 2 (June 1999): 137–50.

4. It was suggested that totipotent cells might be removed from four- or eight-cell, preimplantation embryos destined for in vitro fertilization (without destroying the embryo, a technology performed with some frequency in contemporary reproductive therapeutic settings for purposes of preimplantation genetic diagnosis); see Ronald M. Green, *The Human Embryo Research Debate: Bioethics in the Vortex of Controversy* (New York: Oxford University Press, 2001). It was also suggested that scientists who conduct embryonic cell research ought not themselves be engaged in the destruction of embryos, or, perhaps more correctly, that the activity of embryonic cell research could or should be viewed as morally distinct from that of obtaining cells through the destruction of embryos; see Helga Kuhse and Peter Singer "Individuals, Humans, and Personhood: The Issue of Moral Status" in *Embryo Experimentation: Ethical, Legal, and Social Issues*, ed. Peter Singer, et al. (Cambridge: Cambridge University Press, 1990).

5. "The Decision on Stem Cell Research Excerpts," *Boston Globe*, 10 August 2001.

6. T. Friend, "Half of Stem Cell Money Could Go to Royalties," *USA Today*, 13 August 2001.

7. Committee Hearing of Health and Human Services on Stem Cell Research, 1 August 2001.

8. Jessica Reaves, "The Great Debate Over Stem Cell Research," *Time*, 8 November 2001.

9. Thomas Dennelly, "Using Embryonic Stem Cells in Research is a Crime," *Newsday*, 18 October 2001.

10. D. Peter Snustad and Michael J. Simmons, *Principles of Genetics* (New York: John Wiley and Sons, 2000).

11. "Severe Combined Immunodeficiency: Gene Therapy Licensed in France," *Genomics and Genetics Weekly*, 21 September 2001.

12. *Ethical Issues in Human Stem Cell Research*, a report by the National Bioethics Advisory Commission, Harold T. Shapiro, chair. September 1999, http://bioethics.georgetown.edu/nbac/pubs.html.

13. It was not possible, however, to determine conclusively whether any patents are pending that would be assigned to these organizations, or whether employees, board members, consultants, or others affiliated with these organizations in fact hold patents that are not yet listed as assigned to these organizations but are nonetheless already the property of them.

14. Committee Hearing of Health and Human Services on Stem Cell Research, 1 August 2001.

15. Marker SSEA-1 is negative for Thomson versus positive for Gearhart. Further and much more elaborate descriptions of the cells in the claims sections of each grant, though, could easy be said to obviate any importance of the difference at the SSEA-1 marker.

16. Glenn McGee, "Creative Ownership: Stem Cells and the Ethics of Intellectual Property," *Penn Center for Bioethics Newsletter*, summer 2002.

17. Rebecca Eisenberg, "How Can You Patent Genes," in this volume, page xx.

18. 710 F2d 329 (7th Cir 1983).

19. R. Weiss, "Cloning Firm is Accused of Ignoring Its Ethics Board," *Washington Post*, 14 July 2001, A3; S. Stolberg, "Bioethicists Fall Under Familiar Scrutiny," *New York Times*, 2 August 2001, A1; S. Stolberg, "Ought We Do What We Can Do?" *New York Times*, 12 August 2001, Week in Review; A. Knox, "Ethicist Spurs Debate on Biological Research," *Philadelphia Inquirer*, 17 July 2001, A1; N. Boyce and D. Kaplan, "And Now, Ethics for Sale?" *U.S. News & World Report*, 30 July 2001.

20. Antonio Regalado, "Experiments in Controversy," *Wall Street Journal*, 13 July 2001, B1.

thirteen

INTELLECTUAL PROPERTY AND AGRICULTURAL BIOTECHNOLOGY
Bioprospecting or Biopiracy?

DAVID MAGNUS

There are a number of concerns raised by the patenting of living organisms and germ plasm. These range from the metaphysical (what counts as part of nature versus as invention) to practical policy concerns. Among the primary considerations are worries over justice, particularly with respect to "bioprospecting" or "biopiracy"—the development and patenting of material derived from resources and knowledge in less developed nations for the benefit of corporations based in developed nations. I will briefly survey some of the general concerns before focusing on the issue of biopiracy and the justice arguments they present.

The 1930 U.S. law governing plant patents resulted in the patenting of (asexual) plant organisms. The plant variety protection system (PVP) set up in the United States in the early 1970s created a broader system of protection, by including seeds. The 1980 *Chakrabarty* decision by the U.S. Supreme Court made it possible to patent living organisms (bacteria) and opened the door to the patenting of other organisms, including mammals. Beginning in 1985 (*Ex Parte Hibberd*), it was possible to obtain utility patents on plants as just another kind of invention. The legitimacy of these plant patents has recently been challenged (*JEM Ag Supply* v. *Pioneer Hi-Bred*) on the grounds that the PVP system is meant to handle plant property rights, rather than utility patents. Potentially this could lead to a revisiting of the issues raised in *Chakrabarty*. (See chapters by Chakrabarty, Wilson, and Seide and Stephens on these issues).

The patenting of organisms and their germ plasm raises a host of broad issues. First, there is the question of the "product of nature doctrine." In case law, it has been established that it is not possible to patent "laws of nature" or physical phenomena. The question that then arises is how it is possible to patent an organism. How much of a change in an organism needs to be made before it is considered no longer part of nature, but a result of "being touched by the hand of man?" As Wilson's chapter makes clear, this is a complicated set of issues—engineered organisms seem to represent a middle ground between "naturally occurring" and "manufactured," between "invention" and "discovery." Unsurprisingly, when legal and policy issues hinge on the answer to metaphysical questions, controversy ensues. In such a situation, most positions are defensible, either in favor or against the "ownership of life."

A second objection to agricultural patents is the "common heritage" argument. This objection states that the germ plasm and various organisms (even modified forms) are all essentially a function of the natural world that we all equally inhabit and have inherited. Therefore, we all have a right to share in the benefits of that inheritance. Any intellectual property rights restrict access and use of our shared heritage. In the end, this objection reduces to a version of the previous argument. If a plant variety, a genetically modified organism (GMO), or a gene is a product of nature, it is legitimate to think it is part of our common heritage. If it were truly an innovation or an invention, then it would not really be part of our common heritage. Furthermore, the key justification for a patent system is to promote the general good, through encouragement of both investment (in research and development) and disclosure. Thus, even if plants and other purportedly patentable material were part of our common heritage, there is no reason why patent protection or other intellectual property (IP) regimes could not be enacted as the best means of utilizing that heritage for the general good.

A third general argument against these patents is that they lead to the commodification of life (see essays by Ossorio and Hanson). Patenting of living things requires that we conceive of them in market terms. It implies both ownership and an instrumentalism that is incompatible with many views about the nature of life. This objection has particularly been made from a theological perspective: as a gift from God, life is transformed into an "invention" to be owned and used. It shows a lack of respect and hubris with respect to the world and our relationship to it. This is often expressed in visceral terms as a concern over "playing God." Though this objection has some weight, it is counterbalanced by arguments about the benefits of utilizing the patent system as a way of encouraging the development of products and even organisms that will be helpful to humanity. Every religious tradition recognizes the importance of balancing the need to "pre-serve the garden" with the need to "tend the garden." The key question is whether patenting of organisms, genes, cells, and the like, pushes the appropriate balance to an excessively instrumentalist view of nature.

These arguments and the legal context in which they take place are well chronicled in earlier chapters of this book.[1] But, they also occur in a changing international context. In the rest of this chapter, I will look at some of the ethical and social issues raised by the "ownership of life" in the agricultural context, particularly with respect to less developed coun-tries (LDCs). Until fairly recently, patenting practices varied widely from country to country and practices that were common in the United States and other Northern nations were often restricted or forbidden in the Southern LDCs. The United States and European Union (EU) were often frustrated in their attempts to enforce intellectual property rights in LDCs. Many of these nations only allowed process patents, not product patents. Therefore, generic companies operating in these countries only had to produce the same pharmaceutical product in a different way to circumvent the patent a company had on a drug. Different countries also recognized patents for varying lengths of time. India, for example, only recognized process patents, and only for a period of five years. Many of the Northern pharmaceutical companies found this a problematic situation for their research and development efforts. Pfizer found that prior to government approval of the antiarthritic drug Feldene, a generic competitor already existed in Argentina, and by the time they went to market, they faced com-petition from six generic drug companies. Attempts to change the system to make it more uniform across national boundaries met with failure. The World Intellectual Property Organization (WIPO) was set up as a United Nations agency in 1967 to administer international agreements and treaties with respect to IP issues. The Paris Convention of 1883 required that each nation grant the same patent protection to people from other countries that they grant to their citizens. This requirement did nothing to stop countries from having IP systems that differed markedly from the U.S. system, as long as they were consistent. Therefore, Pfizer and other com-panies began to lobby the WIPO to change the Paris Convention.[2] This effort met with failure. Later, the United States, the EU, and Japan agreed to pursue an alternative. There began to be an increasing connection between IP and trade. This resulted in increasing pressure from the United

States and other Northern nations on the LDCs to comply with their patent protection systems or face trade sanctions. And, because of negotiations over the General Agreement on Tariffs and Trade (GATT), there emerged a "floor" governing IP systems in all GATT nations, through the Trade Related Aspects of Intellectual Property Rights (TRIPS).

The question that the current system raises is whether it is fundamentally unjust. LDCs are systematically disadvantaged relative to the interests of the United States, the EU, and their multinational corporations. The traditional knowledge and the germ plasm of LDCs are mined for their value for industrial interests, often with little or no payback to the original developers of the material. This is sometimes referred to as "biopiracy." There are at least two different arguments. First, LDCs may be responsible for both the creation and the preservation of valuable germ plasm. These organisms have often resulted from years of agricultural practices (similar in many respects to scientific plant production), and efforts by indigenous groups to preserve valuable and rare resources would seem to entitle the developers to some of the benefits that may accrue as the result of usage of the organisms or genes that they have helped to create and preserve.[3] I will refer to this as "resource biopiracy." Second, traditional knowledge also involves knowledge of how the raw materials can be harnessed for various purposes: medicinal, agricultural, and so on. The argument that is made here is that IP built on the basis of traditional knowledge should either entitle the communities that created that knowledge with a share of the benefits that ensue, or more typically that the traditional knowledge constitutes prior art and thus invalidates any IP claims. I will refer to this as "knowledge biopiracy." It is important to recognize that these two arguments may conflict—one is aimed primarily at a share of the benefits of the products that eventually result, while the other attempts to invalidate the legitimacy of the patent claims, leaving development outside of the IP system. I will discuss several examples to illustrate the issues at stake in what are often treated as paradigm cases of biopiracy.

THE BEAN WARS

According to his patent application, in 1994 Larry Proctor purchased a bag of assorted beans in Mexico. He selected the yellow beans and brought them back to the United States to try growing them in Montrose County, Colorado. He began crossbreeding and selecting and found that he had a plant that had many desirable characteristics: heartier, more moisture resistant pods, and a distinctive yellow color. In 1996 Proctor applied for patent protection and in 1999 received both a utility patent and a U.S. Plant Variety Protection certificate. This patent covers any yellow beans (of a certain shade) from *Phaseolus vulgaris*.

The yellow beans that Proctor found in his "package of dry edible beans" in Mexico were not unique. Mexican breeders have been growing yellow beans for centuries. In recent years, agronomists have been crossing them and producing improved varieties. In 1978 the Mayacoba bean was developed, and has since become quite popular in parts of northern Mexico. There are several other yellow varieties that are also common. In 1994 Rebecca Gilliland began working with a bean cooperative in Los Mochas to arrange to export the Mayacoba to the United States Exports gradually increased until 1999 when Proctor's company, Pod-NERS, claimed that Gilliland was infringing on his patent. He demanded six cents per pound (the beans now sell for roughly twenty-seven cents per pound) to license the selling of the beans. Gilliland refused, and she was subsequently served with a patent complaint. She claims that Proctor demanded that U.S. customs agents inspect her produce and bean shipments at the border to prevent yellow beans from being brought into the United States. The result is that the beans are no longer exported, and the farming cooperative has lost out on investments in sorters and stoners.

This case has enraged both opponents of the practice of patenting genes and organisms as well as those who advocate better use of the IP system in LDCs. The Mexican government filed a suit to challenge the patent claims, and more recently, the Center for International Tropical Agriculture (ICTA) has also filed a claim against Proctor's patent. Under terms of the 1994 agreement between the Consultative Group on International Agricultural Research (CGIAR) and the United Nations Food and Agriculture Organization, any germ plasm maintained by the CGIAR is part of the public domain. Intellectual property claims can not be made on any of this material. There are several varieties of yellow beans in the CGIAR's holdings that would seem to infringe the Proctor patent. There are reports that genetic analysis performed by Mexico's National Research Institute for Agriculture, Forestry, and Livestock as well as by the ICTA have demonstrated that Proctor's Enola bean is genetically identical to plants in the CGIAR holdings (though interestingly, not to the Mayacoba). Further, many geneticists have argued that the two years between 1994 (when the patent claims Proctor found the beans) and 1996 (the time of filing) are not sufficient to truly develop a novel plant variety. In response, Proctor now claims he found the initial yellowish beans in Mexico in 1990, rather than in 1994 as he had earlier claimed.

The case of the Enola bean illustrates beautifully the concept of resource biopiracy. There is no disputing that the raw material for the variety produced by Larry Proctor came from Mexico. It is also clear that the beans that he used (or similar beans) were a product of efforts to both produce and preserve particular varieties, and that ongoing efforts by local farmers and agronomists were underway to produce similar results (similar enough to violate the patent that was issued). Yet those efforts did not result in any IP claims on the part of the developers of the Mayacoba or any of the many other yellow varieties of *Phaseolus vulgaris*. And to answer the claims of Proctor, the CGIAR must undertake expensive litigation. Moreover, success hinges on establishing that the prior art invalidates the claim. In the absence of a native IP culture to protect the interests of local farmers, it is imperative that there be seed banks that can compare older samples with putative new varieties as a way of establishing prior art.

TURMERIC

Turmeric has been widely utilized for healing wounds by people in India for centuries. In 1995 two University of Mississippi researchers (of Indian descent), Drs. Soman K. Das and Hari Har P. Cohly, were granted a patent on the use of a certain amount of turmeric as effective for healing wounds. Their patent on "an effective amount of turmeric powder" included both the oral and the topical application of the treatment for surgical wounds and body ulcers. The Council of Scientific and Industrial Research (CSIR) later challenged this patent. A good deal of documentary evidence established the long-standing use of turmeric for precisely these purposes. All told, thirty-two documents were submitted from a large literature on the subject. The patent was canceled in 1997. This case was widely hailed as both a triumph of the existing regulatory system and a demonstration of the inadequacy of the U.S. patent system to prevent biopiracy.

This is a clear example of knowledge biopiracy. As the subsequent challenge clearly demonstrates, the use of turmeric for medical purposes has a very long history. However, it took an expensive legal challenge to successfully overturn the patent. And the key to success was the production of documentary evidence to establish that the use of turmeric for treating of wounds constitutes "prior art." This means that cultures whose practices are not documented will not be recognized, even if they are quite common and well known locally.

NEEM

The neem tree (a member of the mahogany family) has a variety of uses in India, ranging from medicinal properties to the use of an extract derived from the neem that has proved effective as an insecticide. In addition to its well-established properties as an insecticide, its twigs have been used to clean teeth, its leaves have been used to brew tea to treat a range of ailments, and juice from its leaves has been used to treat skin disorders. The neem tree has been referred to as "the village pharmacy" for all of its uses.[4] Researchers studying many of these properties have attempted to develop products out of the neem. Over thirty patents have been granted on the neem, including one by W. R. Grace of Boca Raton, Florida, on a process for fractionating oils of the neem so that they are more stable and hence can be stored for much longer than the oil extracts that are commonly used as an insecticide. Grace claims the shelf life has been extended from a few days to two years. This patent was granted in 1992, but was challenged by the Foundation on Economic Trends. The challenge was denied, largely due to a failure to offer substantial documentation. Grace holds a number of other patents as well, and at least one them—a fungicide derived from the neem seeds— was finally invalidated (after a lengthy legal and political battle) by the European Patent Office in 2000 on the grounds that the process for which the patent was obtained was actually demonstrably in use for some time. This case has produced a tremendous controversy. Regardless of the substantive legal issues, the neem tree has a social and cultural meaning which makes the prospect of "owning it" far more problematic than any straightforward legal analysis would indicate.

This is in many ways the most complex case. The claim on neem for a pesticide represents both resource and knowledge biopiracy and was overturned. More complex is the process patent procured by Grace for developing a more stable variety of the pesticide. This would seem (unlike the previous two examples) to be a fairly straightforward example of the value of bioprospecting. However, it is important to recognize that the material came from India, the basic knowledge of the properties of the neem products and its uses came from India, and there was an indigenous ongoing research program from the 1920s on attempting to derive products from neem including more stable versions of the pesticide. It is unsurprising that developed nations and their corporations would succeed more quickly than the basic research scientists within the LDCs. The current patent system rewards only the conventionally defined "winners" of the race—and this means that the relative latecomers to the research (such as Grace) can build on what has gone before and still procure the patents and de facto win control over future uses of the neem as a pesticide. Thus, the nations that produced the germ plasm and the knowledge that made the developments possible are left without any recognized stake in the eventual rewards that issue from the technology.

The turmeric case is a wonderful example of knowledge biopiracy. Again, the key to the invalidation of the claim was the documentation that established that the usage of turmeric had already been well established prior to the patent issued to the University of Mississippi researchers.

ISSUES RAISED BY THESE CASES

There are several ethical problems with the current IP system as it stands. First, with respect to germ plasm from LDCs, critics have pointed out the disparity between the way genetic resources and other natural resources are treated. Petroleum or mineral resources are the property of the nation within which they reside. Genetic resources, in contrast, "have long been considered a common heritage available to other nations for free."[5] As the case of the "Bean Wars" indicates, the LDCs often act as stewards of these resources and even utilize scientific methods to develop them. But there is no systematic way of rewarding them for preserving and developing these resources. One solution to this problem has been to create the Convention on Biological Diversity (CBD). This requires that researchers from signatory nations must seek permission prior to utilizing genetic resources from other countries. However, given the extent to which valuable genes and organisms have already flowed from LDCs to the North, and the additional problems of enforcement, it is not clear that the CBD is a panacea for these problems. Further, it is not clear how to address the conflict between the CBD and the TRIPS.[6] From a moral point of view, the fact that local communities preserve and create genes and organisms would seem to require that ethically they are entitled to some form of benefit sharing in the fruits of future product developments based on their material. The CBD is one mechanism that can help promote benefit sharing, but more needs to be done.

Second, there is the way that the regulatory system seems stacked against LDCs. As the neem and turmeric cases in particular make clear, a U.S. patent can only be invalidated by claims of prior use if there is evidence of use in a form recognizable by the U.S. courts and patent and trademark office examiners. That is, accessible documentation or prior patents are needed to invalidate a claim. If indigenous, traditional knowledge is largely expressed in customs, habits, and oral traditions, they will not be recognized in the patent system. There may even be problems if the documentation is not in English or an easily accessible language. This results in systematically favoring nations with well-established IP systems similar to the one in the United States, and it favors nations that have strong written rather than oral traditions.

In the area of biotech patents, particularly in the United States, there seems to be a great deal of leniency in granting patents by examiners and reliance on the courts and other systems of appeal to overturn patents that are invalid. The problem with this is that it favors nations and corporations with financial resources, which have an incentive to attempt to procure as many patents as possible, whether valid or not, and place the burden on poor LDCs or nongovernment organizations (NGOs) to attempt to invalidate them. Invalidating the turmeric patent cost several hundred thousand dollars, as will the attempt to invalidate the Enola bean patent. It is simply not possible for every one of the many alleged cases of biopiracy to be challenged by the LDCs or the NGO community. It is important to note that the previous argument (which applied to resource biopiracy) leads to a very different conclusion than this argument (which will apply primarily to knowledge biopiracy). This argument challenges the validity of the patents at all, rather than making a case for benefit sharing.

The ramping up of the IP systems in some of the LDCs might be helpful in trying to deal with some of these problems.[7] Peter Drahos has suggested creating a Global Bio-Collecting Society to better reward the contributions of indigenous groups for their knowledge.[8]

A third argument against the current system is that the LDCs are arguably deprived of future potential benefits through the opportunity to develop products based on their indigenous genetic resources and/or traditional knowledge. For example, if a company such as W. R. Grace has, in fact, substantially created an innovative process to create a more stable version of the neem pesticide, the current patent system makes them the sole beneficiary. However, it is quite possible that local industries would have eventually produced the same product. Indeed, native neem researchers have conducted most of the research done on the neem tree, beginning in the 1920s. They were ignored for decades. This pits the interests of developed nations, which want products to be developed as quickly as possible, against the interests of the LDCs. Again, this system will favor those nations and institutions (corporations and universities) with the resources to develop products as quickly as possible, building on the genetic resources and traditional knowledge of LDCs.[9] This is particularly problematic because, in addition to losing future economic opportunities, it may undercut current local businesses as new products undercut existing ones. For example, current neem-based pesticides (with a short shelf life) will presumably lose out in the marketplace in competition with Grace's more stable neem-based product. The loss of traditional industries in competition with superior products could result in food security problems in LDCs.

This argument underscores the importance of developing ways of sharing the benefits with the communities that materially contribute to the development of valuable IP, not just those eventually defined as the "inventors."[10]

Finally, we need to be concerned about the impact of bioprospecting and biopiracy on the lives of the people in LDCs. When Grace processes twenty tons of neem seed per day in India, it changes the agricultural practices in many ways. It drives up the price of seeds and creates jobs with relatively high wages at the same time that it destroys other jobs. Consideration of these impacts, for better and for worse, must be thought about as part of the consideration of the impact of bioprospecting. If the net harms in a particular case outweigh the benefits (for the LDC), there must be mechanisms that can protect the use of their germ plasm, and their knowledge, so that it does not harm them.

Justice requires attention to more than the technical elements of current patent law. It is much more than a technical area of law, and the assessment of what institutions, laws, and practices we should adopt has to involve more than a scientific assessment of the risks and benefits of various IP regimes. Who benefits? Who has a claim on that benefit? Who is exposed to risk? and Who decides the allocation of risk and benefit? are all moral questions that require much more serious attention.

NOTES

1. For more on these and the biopiracy arguments, see Donald Bruce and Ann Bruce, *Engineering Genesis* (London: Earthscan, 1998).

2. Peter Drahos, "Global Property Rights in Information: The Story of TRIPS at the GATT," *Prometheus* 13 (1995): 6–19 and Copyright 1995 by the president and fellows of Harvard College. Harvard Business School Case 9-392-073. This case was prepared by Michael A. Santoro under the supervision of Professor Lynn Sharp Paine as the basis for class discussion rather than to illustrate either effective or ineffective handling of an administrative situation.

3. Anil K. Gupta, "Scientific Perception of Farmers Innovations in Dry Regions: Barriers to the Scientific Curiosity," published under the title "Scientists' View of Farmers' Practice in India: Barriers Interaction" in *Farmer First: Farmer Innovation and Agricultural Research*, ed. Robert Chambers, Arnold Pacey, and Lori Ann Thrupp (London: Intermediate Technology, 1989), 24–30.

4. Boyce Rensberger, "Is Any Tree As Useful As a Neem?" *Washington Post*, 17 February 1992, A3.

5. Ashok Sherma, "Tree Focuses Debate on Control of Resources," *Los Angeles Times*, 19 November 1995.

6. Michael Blakeney, Joel Cohen, and Stephen Crespi, "Intellectual Property Rights and Agricultural Biotechnology" in *Managing Agricultural Biotechnology, Addressing Research Program Needs and Policy Implications*, ed. Joel Cohen (The Hague: CABI Publishing and the International Service for National Agricultural Research, 1999).

7. Calestous Juma, *The Gene Hunters* (Princeton: Princeton University Press, 1989); Anil K. Gupta, "Rewarding Creativity for Conserving Diversity in Third World: Can IPR Regime Serve the Needs of Contemporary and Traditional Knowledge Experts and Communities in Third World?" in *Strategic Issues of Industrial Property Management in a Globalising Economy*, APPI Forum Series, ed. Thomas Cottier, Peter Widmer and Katharina Schindler (Oxford: Hart Publishing, 1999), 119–29.

8. Peter Drahos, "Indigenous Knowledge, Intellectual Property and Biopiracy: Is a Global Bio-Collecting Society the Answer?" *European Intellectual Property Review* 22 (2002): 245–50.

9. Paul Thompson, *Food Biotechnology is Ethical Perspective* (London: Blackie Academic, 1997), 163 ff.

10. Jon Merz, this volume.

glossary

allele
A variant of a single gene, inherited at a particular genetic locus.

amino acids
The molecular building blocks of proteins; a protein is a chair of amino acids in a certain sequence. There are twenty main amino acids, and their order determines the function of the protein they create.

anneal
The process through which complementary single strands of DNA "recognize" each other and fuse together to form a double-stranded unit.

art or prior art
A term used in consideration of the problem of patentable novelty, encompassing all that is known prior to the filing date of the application in the particular field of the invention, represented by already issued patents and publications.

artificial selection
Selective breeding, carried out by humans, to alter a population, most frequently used to increase the frequency of a desired trait.

bacteriophage
A virus that infects and replicates exclusively within a bacterium, usually killing it.

cDNA   Strong, cloned copies of otherwise fragile mRNA—the essential messenger element of the genes in the DNA which help in the coding of proteins.

chromosome   Structure in the cell nucleus that carries the DNA. At certain times in the cell cycle, they are visible as string-like entities. The number of chromosomes varies from species to species. The number found in human cells is forty-six (twenty-three pairs), half of which are inherited from each parent.

cloning   The process of molecular cloning involves isolating a DNA sequence of interest and obtaining multiple copies of it in an organism, usually a bacterium, that is capable of growth over extended periods. Large quantities of the DNA molecule can be then isolated in pure form for detailed molecular analysis. The ability to generate virtually endless copies (clones) of a particular sequence is the basis of recombinant DNA technology and its application to human and medical genetics.

codon   A triplet of nucleotides in the DNA coding for one amino acid. The sequence of codons in DNA determines the sequence of amino acids in proteins, and therefore the structure of the protein being made.

complementarity   In molecular biology, the relationship of the nucleotide bases on two different strands of DNA or RNA. When the bases are properly paired—(adenine with thymine (DNA) or uracil (RNA)—the strands are complementary.

differentiation   The structural and functional modification of an unspecialized cell into a specialized one.

DNA   Deoxyribonucleic acid; the molecule that controls inheritance.

eugenics   A branch of science that involves using principles of genetics to "improve" humankind. Though presently out of favor, the idea that this was a good thing was fairly universally accepted throughout the early part of the twentieth century.

exon   The nucleotide sequences of some genes consist of parts that code for amino acids, with other parts that do not code for amino acids interspersed among them. The coding parts, which are translated, are called exons; the interspersed noncoding parts are called introns.

expressed sequence tags (ESTs)   A small part of the active part of a gene which can be used to fish the rest of the gene out of the chromosome. While the EST itself is not "functional" (it does not code for a protein), many researchers are attempting to obtain patents on them. Opponents of gene patenting argue that since the functions of ESTs are not known, they fail the requirements for patentable material.

gene   A sequence of nucleotides coding for a protein (or part of a protein).

gene fragments   Pieces of genes containing only the exons (those parts of the gene which actually encode the protein sequence). They are composed of cDNA.

gene pool   All the genes in a population at a particular time.

genome   The full set of DNA in a cell or organism.

genotype   The set of two genes at a locus possessed by an individual.

germ plasm   The reproductive cells in an organism; the cells that produce the gametes (sperm and egg). All the cells in an organism can be divided into the soma and the germ cells.

germline engineering A process that involves making "improvements"
or enhancement   in the gametes (sperm or egg) of an organism. These changes will be passed on to subsequent generations.

hybrid   Offspring of a cross between two species.

hybridization   The interaction of complementary nucleic acid strands. Since DNA is a double-stranded structure held together by complementary interactions (in which the nucleotide cytosine always binds to guanine, and adenine to thymine), complementary strands favorably reanneal (join together) or "hybridize" to each other when separated.

intron   *See* **exon**.

linked   Referring to genes located on the same chromosome.

locus   The location in the DNA occupied by a particular gene.

mutation   A random change in the sequence of nucleotides of the chromosome.

nonobvious

In order for a patent to be granted, the claimed invention must be "nonobvious" to one of "ordinary skill in the art." In other words, if one obtains a new and unexpected result, the invention is said to be nonobvious.

novelty

A requirement for patentability. If an invention has been used or was known to others, it is probably no longer novel and therefore not eligible for patent protection.

nucleotide

The building block of DNA. There are four basic building blocks, which are arranged in units of three called codons.

oligonucleotides

A short polymer of, for example, twenty or so deoxyribonucleotides or ribonucleotides; thus a fragment of DNA or RNA.

operon

A set of functionally unified structural genes and the regulating genes that control them.

patent

A grant issued in the name of the United States under the seal of the Patent and Trademark Office (PTO), which "confers the right to an applicant to exclude others from making, using, or selling an invention in the United States" and its territories for twenty years from the application filing date.

phenotype

The outward appearance or expression of an organism's genotype.

plasmid

A small circular form of DNA found in bacteria that carries certain genes, such as for antibiotic resistance, and that replicates independently of the host cell.

plant patents   Grants issued to any person who has invented or
                           discovered and asexually reproduced any distinct
                           and new variety of plant, including cultivated
                           spores, seeds, mutants, hybrids, and newly found
                           seedlings, other than a tuber-propagated plant or
                           a plant found in an uncultivated state. A plant
                           patent has a term of twenty years from the appli-
                           cation date.

protein   A molecule made up of a sequence of amino
                           acids. Proteins are the most common organic
                           molecule found in living organisms.

recombinant DNA   DNA which has been altered by joining genetic
                           material from two different sources. It usually
                           involves putting a gene from one organism into
                           the genome of a different organism, generally of
                           a different species.

restriction   An enzyme that will recognize, bind to, and
endonuclease   hydrolyze (break apart) specific nucleic acid
                           sequences in double-stranded DNA.

soma   The mortal cell lines in a body. *See* **germ plasm**.

terminator gene   A gene specifically inserted into a plant which
                           programs the plant's seeds to sterilize themselves
                           by destroying their own embryos.

reverse transcriptase An enzyme that catalyzes the synthesis of DNA
                           from using RNA as a template.

transposable   Units of DNA that move from one DNA mole-
elements, or   cule to another, inserting themselves at random.
transposons   Can also catalyze DNA rearrangements (muta-
                           tions) such as deletions and inversions.

utility patents   Grants issued to anyone who invents or discovers a new and useful process, machine, manufacture, or compositions of matter, or any new and useful improvement thereof. A utility patent has a term of twenty years from the application date.

vector   An agent, often a virus or plasmid, used to carry foreign DNA into a cell.

contributors

ANANDA CHAKRABARTY, Ph.D., is a distinguished professor in the Department of Microbiology and Immunology at the University of Illinois College of Medicine. He pioneered the use of genetic engineering in his creation of a new life-form, a bacterium with the ability to break down crude oil. His litigious battle for patent protection resulted in the Supreme Court decision that paved the way for future patenting of biotechnological discovery.

JACK WILSON, Ph.D., is an associate professor of philosophy at Washington and Lee University. Formerly a fellow in the History and Philosophy of Science Program at Northwestern University, he is the recent author of *Biological Individuality: The Identity and Persistence of Living Entities.*

ROCHELLE K. SEIDE, M.S., J.D., Ph.D., is a partner in the Intellectual Property Department of Baker Botts, LLP. She practices as a patent attorney with technical expertise in biotechnology and the pharmaceutical/chemical arts. Formerly a professor and researcher of medical genetics and microbiology, she is currently a chair of the Education/Biotechnology Board Subcommittee of the American Intellectual Property Law Association and the AIPLA representative for the U.S. Patent and Trademark Office Biotechnology Customer Partnership.

DANIEL KEVLES, Ph.D., directs the Program in Science, Ethics, and Public Policy at the California Institute of Technology where he is the Koepfli Professor of Humanities. A historian of science and society, he has authored many books and articles on genetics and advancing technology including the award-winning *The Baltimore Case: A Trial of Politics, Science, and Character* and *In the Name of Eugenics: Genetics and the Uses of Human Heredity*.

ROBERT LEE HOTZ, M.A., is a Pulitzer Prize–winning science and technology writer for the *Los Angeles Times*. Widely recognized for his achievements in science journalism, he has authored *Designs On Life: Exploring The New Frontier of Human Fertility*, which examines the scientific, legal, and political issues posed by human embryo experiments and the commercial conception industry.

LORI B. ANDREWS, J.D., is the director of the Institute for Science, Law, and Technology at Illinois Institute of Technology and a professor of law at Chicago-Kent College of Law. She has served on numerous federal advisory committees and was chair of the Working Group on the Ethical, Legal, and Social Implications of the Human Genome Project. She is the author of seven books and over one hundred articles on biotechnological issues including genetics, reproductive technologies, and cloning. Her most recent book is *The Clone Age: Adventures in the New World of Reproductive Technology*.

DOROTHY NELKIN holds a university professorship at New York University in the Department of Sociology and the School of Law. She is the author of many books and articles on genetics and biotechnology, including (with Susan Lindee) *The DNA Mystique: The Gene as Cultural Icon* and (with Lori Andrews) *Body Bazaar: The Market for Human Tissue in the Biotechnology Age*.

PILAR N. OSSORIO, J.D., Ph.D., is an assistant professor of law and medical ethics at the University of Wisconsin-Madison. She was formerly the director of the Genetics Section of the American Medical Association's Institute for Ethics. She is the author of numerous publications on topics including the social and ethical issues of biotechnology patenting and the roles of race, gender, and access in the future of genetic medicine.

JON F. MERZ, J.D., M.B.A., Ph.D., is an assistant professor of bioethics in the Department of Molecular and Cellular Engineering and faculty associate in the Center for Bioethics at the University of Pennsylvania. He has published extensively on issues including genetic technologies, research ethics and regulation, intellectual property, informed consent, reproductive rights and policy, and privacy and confidentiality in medicine and research.

REBECCA S. EISENBERG, J.D., is a professor of law at the university of Michigan Law School and a visiting professor at Stanford Law School. She has written extensively about patent law as applied to biotechnology and the role of intellectual property at the public–private divide in research science. She has served as chair of intellectual property rights working groups for the National Academy of Sciences and the National Institutes of Health.

MARK HANSON, M.A.R., Ph.D., is a research professor at the University of Montana's Practical Ethics Center and is executive director of the Missoula Demonstration Project. He was previously the director of the Hastings Center research program on Values and Biotechnology. He has published widely on the moral and religious values challenged by gene patenting.

DAVID B. RESNIK, M.A., Ph.D., is an associate professor of medical humanities at the Brody School of Medicine at East Carolina University and associate director of the Bioethics Center at University Health Systems of Eastern North Carolina. He has published over forty articles on various topics in the philosophy of biology and medicine, and bioethics and is the author of *The Ethics of Science: An Introduction* and *Human Germline Gene Therapy: Scientific, Moral, and Political Issues.*

DAVID MAGNUS, Ph.D., is the graduate studies director and assistant professor at the Center for Bioethics at the University of Pennsylvania. He has published articles on topics in the philosophy and history of biology and on a range of bioethical issues in genetics, research ethics, and reproductive medicine. He was the director of an NEH–NSF summer institute for college and university faculty on "Scientific, Ethical, and Social Consequences of Genetic Technology." He serves as associate editor of the *American Journal of Bioethics*. He has served as a consultant for the National Conference of State Legislators and for the World Bank.

GLENN MCGEE, Ph.D., is assistant professor and associate director for education at the Center for Bioethics at the University of Pennsylvania. He is editor in chief of the *American Journal of Bioethics*. The author of numerous publications on bioethics, particularly on issues of reproduction and genetics, he is the author of *The Perfect Baby*, and the editor of *The Cloning Debate* and *Pragmatic Bioethics*. His work has focused on introducing the methods of American pragmatism to various debates within bioethics, ranging from genetic testing, eugenics, gene therapy, and egg donation to issues in research and clinical ethics.

ART CAPLAN is trustee professor of bioethics and director of the Center for Bioethics at the University of Pennsylvania. The author of over four hundred articles, he has edited or authored over a dozen books. His work spans the entire range of issues in bioethics, including genetics and gene patenting. His most recent books include *Am I My Brother's Keeper*, *Due Consideration*, *The Ethics of Organ Transplants* (with Dan Coehlo), and *Assisted Suicide* (with Lois Snyder).

index